Essential Forensic Biology

Alan Gunn
Liverpool John Moores University

John Wiley & Sons, Ltd

Other Wiley Editorial Offices

John Wiley & Sons Inc., 111 River Street, Hoboken, NJ 07030, USA

Jossey-Bass, 989 Market Street, San Francisco, CA 94103-1741, USA

Wiley-VCH Verlag GmbH, Boschstr. 12, D-69469 Weinheim, Germany

John Wiley & Sons Australia Ltd, 33 Park Road, Milton, Queensland 4064, Australia

John Wiley & Sons (Asia) Pte Ltd, 2 Clementi Loop #02-01, Jin Xing Distripark, Singapore 129809

John Wiley & Sons Canada Ltd, 22 Worcester Road, Etobicoke, Ontario, Canada M9W 1L1

Wiley also publishes its books in a variety of electronic formats. Some content that appears in print
may not be available in electronic books.

British Library Cataloguing in Publication Data

A catalogue record for this book is available from the British Library

ISBN-13 978-0470-01276-5 (HB)   ISBN-10 0470-01276-5 (HB)
ISBN-13 978-0470-01277-2 (PB)   ISBN-10 0470-01277-3 (PB)

Typeset in $10^1/_2/12^1/_2$ Sabon by SNP Best-set Typesetter Ltd., Hong Kong
Printed and bound in Great Britain by Antony Rowe Ltd., Chippenham, Wilts
This book is printed on acid-free paper responsibly manufactured from sustainable forestry in which
at least two trees are planted for each one used for paper production.

To Sarah, who believes that no evidence is required in order to find a husband guilty.

Thanks to Sarah and to all of the academic and technical staff at the School of Biological & Earth Sciences, John Moores University who helped me along the way.

Contents

Introduction

The word 'forensic' derives from the Latin *forum* meaning 'a market place': in Roman times this was where business transactions and some legal proceedings were conducted. For many years the term 'forensic' had a restricted definition and denoted a legal investigation but it is now commonly used for any detailed analysis of past events, i.e. when one looks for evidence. For example, tracing the source of a pollution incident is now sometimes referred to as a 'forensic environmental analysis', determining past planetary configurations is referred to as 'forensic astronomy', whilst historians are said to examine documents in 'forensic detail'. For the purposes of this book, 'forensic biology' is defined broadly as 'the application of the science of biology to legal investigations' and therefore covers human anatomy and physiology, organisms ranging from viruses to vertebrates and topics from murder to the trade in protected plant species.

Although forensic medicine and forensic science only became specialized areas of study within the last 200 or so years, their origins can be traced back to the earliest civilizations. The first person in recorded history to have medico-legal responsibilities was Imhotep, Grand Vizier, Chief Justice, architect and personal physician to the Egyptian pharaoh Zozer (or Djoser). Zozer reigned during 2668–2649 BC and charged Imhotep with investigating deaths that occurred under suspicious circumstances. The codification of laws was begun by the Sumerian king Ur-Nammu (ca. 2060 BC) with the eponymous 'Ur-Nammu Code' in which the penalties of various crimes were stipulated whilst the first record of a murder trial appears on clay tablets inscribed in 1850 BC at the Babylonian city of Nippur. The word 'murder' probably derives from the Old English 'morder' or 'morther' (verb = *mortheren*) whilst homicide is derived from the Latin '*homo*', a man, and '*caedere*' meaning to strike, cut or kill. A homicide may be either a murder, if malice was involved, or a consequence of manslaughter if it occurred as an accident or during self-defence. Although the ancient Greeks are known to have performed human dissections, Julius Caesar (102/100 – 44 BC) has the dubious distinction of being the first recorded murder victim in history to have undergone an autopsy. After being assassinated, his body was examined by the physician Antistius who concluded that although Julius Caesar had been stabbed 23 times only the second of these blows, struck between the first and second ribs, was fatal.

Essential Forensic Biology, Alan Gunn
© 2006 John Wiley & Sons, Ltd

In England, the office of Coroner dates back to the era of Alfred the Great (871–899) although his precise functions at this time are not known. It was during the reign of Richard I (1189–1199) that the Coroner became an established figure in the legal system. The early Coroners had widespread powers and responsibilities that included the investigation of crimes ranging from burglary to cases of murder and suspicious death. The body of anyone dying unexpectedly had to be preserved for inspection by the Coroner, even if the circumstances were not suspicious. Failure to do so meant that those responsible for the body would be fined, even though it might have putrefied and created a noisome stench by the time he arrived. It was therefore not unusual for unwanted bodies to be dragged away at night to become another village's problem. The Coroner's responsibilities have changed considerably over the centuries but up until 1980 he was still expected to view the body of anyone dying in suspicious circumstances. Although the Coroner was required to observe the corpse he did not undertake an autopsy. In European countries, the dissection of the human body was considered sinful and was banned or permitted only in exceptional circumstances until the nineteenth century. Most Christians believed that the body had to be buried whole otherwise the chances of material resurrection on Judgement Day were slight. The first authorized human dissections took place in 1240 when the Holy Roman Emperor Frederick II decreed that a corpse could be dissected at the University of Naples every five years to provide teaching material for medical students. Subsequently, other countries followed suit, albeit slowly. In 1540, King Henry VIII became the first English monarch to legislate for the provision of human dissections by allowing the Company of Barber Surgeons the corpses of four dead felons per annum, and in 1663 King James II increased this figure to six per annum. Subsequently, after passing the death sentence, judges were given the option of permitting the body of a convicted criminal to be buried (albeit without ceremony) or to be exposed on a gibbet or dissected. Nevertheless, the lack of bodies and an eager market among medical colleges created the trade of body snatching. Body snatchers were usually careful to leave behind the coffin and the burial shroud because taking these would count as a felony (i.e. a serious criminal offence), which was potentially punishable by hanging. Removing a body from its grave was classed as merely a misdemeanour. The modern-day equivalent is the Internet market in human bones of uncertain provenance (Huxley and Finnegan, 2004; Kubiczek and Mellen, 2004).

The first postmortem to determine the cause of a suspicious death took place in Bologna in 1302. A local man called Azzolino collapsed and died suddenly after a meal and his body very quickly became bloated whilst his skin turned olive and then black. Azzolino had many enemies and his family believed that he had been poisoned. A famous surgeon, Bartolomeo de Varignana, was called upon to determine the cause and he was permitted to undertake an autopsy. He concluded that Azzolino had died as a consequence of an accummulation of blood in veins of the liver and that the death was therefore not suspicious.

Although this case set a precedent, there are relatively few records from the following centuries of autopsies being undertaken to determine the cause of death in suspicious circumstances.

The first book on forensic medicine may have been that written by the Chinese physician Hsu Chich-Ts'si in the sixth century AD but this has since been lost. Subsequently, in 1247, the Chinese magistrate Sung Tz'u wrote a treatise entitled '*Xi Yuan Ji Lu*' that is usually translated as 'The Washing Away of Wrongs', and this is generally accepted as being the first forensic textbook (Sung and McKnight, 1981; Peng and Pounder, 1998). Sung Tz'u would also appear to be the first person to apply an understanding of biology to a criminal investigation as he relates how he identified the person guilty of a murder by observing the swarms of flies attracted to the bloodstains on the man's sickle. In Europe, medical knowledge advanced slowly over the centuries and forensic medicine really only started to be identified as a separate branch of medicine in the 1700s (Chapenoire and Benezech, 2003). The French physician Francois-Emanuel Foderé (1764–1835) produced a landmark three-volume publication in 1799 entitled *Les lois éclairées par les sciences physiques: ou Traité de médecine-légale et d'hygiène publique* that is recognized as a major advancement in forensic medicine. In 1802, the first chair in Forensic Medicine in the UK was established at Edinburgh University and in 1821 John Gordon Smith wrote the first book on forensic medicine in the English language entitled '*The Principles of Forensic Medicine*'.

Animals and plants have always played a role in human affairs, quite literally in the case of pubic lice, and have been involved in legal wrangles ever since the first courts were convened. Disputes over ownership, the destruction of crops and the stealing or killing of domestic animals can be found in many of the earliest records. For example, Hammurabi, who reigned over Babylonia during 1792–1750 BC, codified many laws relating to property and injury that subsequently became the basis of Mosaic Law. Amongst these laws it was stated that anyone stealing an animal belonging to a freedman must pay back tenfold, whilst if the animal belonged to the court or a god then he had to pay back thirtyfold. In the absence of a money-based economy, the paying of animals and plants as fines was commonplace whilst humans have always been sadistically inventive in their use of animals as executioners to kill condemned prisoners. The Romans threw Christians (and others) to the lions, whilst in the 1920s and 1930s prisoners in Siberian gulags were tied to posts and left to be exsanguinated by the hoards of mosquitoes (Applebaum, 2004). Animals have also found themselves in the dock accused of various crimes. In the Middle Ages there were several cases in which pigs, donkeys and other animals were executed by the public hangman following their trial for murder or sodomy. The judicial process was considered important and the animals were appointed a lawyer to defend them and they were tried and punished like any human. In 1576, the hangman brought shame on the German town of Schweinfurt by publicly hanging a pig in the custody of the court before due process had taken place. He never worked in the town again and his behaviour is said to have given rise to the term 'Schweinfurter Sauhenker' (Schweinfurt sow hangman) to describe a disreputable scoundrel (Evans, 1906). However, the phrase has now fallen out of fashion. Today, it is the owner of a dangerous animal who is prosecuted when it wounds or kills someone, although it may still find itself facing the death penalty.

During the nineteenth century, a number of French workers made detailed observations on the sequence of invertebrate colonization of human corpses in cemeteries and attempts were made to use this knowledge to determine the time since death in murder investigations (Benecke, 2001). Thereafter, invertebrates were used to provide evidence in a sporadic number of murder investigations but it was not until the 1980s that their potential was widely recognized. Part of the reason for the slow development is the problem of carrying out research that can be applied to real-case situations. The body of the traditional experimental animal, the laboratory rat, bears so little resemblance to that of a human being that it is difficult to draw meaningful comparisons from its decay and colonization by invertebrates. Pigs, and in particular foetal pigs, are therefore the forensic scientists' usual choice of corpse although America (where else?) has a 'Body Farm' in which dead humans can be observed decaying under a variety of 'real life (death?) situations' (Bass and Jefferson, 2003). Leaving any animal to decay inevitably results in a bad smell and attracts flies, so it requires access to land far from human habitation. It also often requires the body to be protected from birds, dogs and rats that could drag it away. Consequently, it is difficult both to obtain meaningful replicates and to leave the bodies in a 'normal' environment. Even more importantly, these types of experiments conflict with European Union Animal By Products Regulations that require the bodies of dead farm and domestic animals to be disposed of appropriately to avoid the spread of disease – and leaving a dead pig to moulder on the ground clearly contravenes these.

The use of animals other than insects in forensic investigations has proceeded more slowly and that of plant-based evidence has been slower still. The first use of pollen analysis in a criminal trial appears to have taken place in 1959 (Erdtman, 1969) and although not widely used in criminal trials since then its potential is now being increasingly recognized (Coyle, 2004). By contrast, the use of plants and other organisms in archaeological investigations has been routine for many years.

The use of molecular biology in forensic science is now well established and it is an accepted procedure for the identification of individuals. This is usually on the basis of DNA recovered from blood and other body fluids or tissues such as bone marrow, and Jobling and Gill (2004) provide a thorough review of current procedures and how things may develop in the future. The use of molecular biology for forensic examination of non-human DNA is less advanced, although this situation will probably improve in the near future as DNA databases become established. When this happens, forensic biology can be expected to play an even larger part in legal proceedings, especially in relation to the illegal trade in protected species, fraud (e.g. false insurance or subsidy claims and mislabelling of food), rustling (horses and farm animals are frequently the target of thieves) and cases of animal cruelty.

One of the major stumbling blocks to the use of biological evidence in English trials is the nature of the legal system (Pamplin, 2004). In a criminal prosecution case, the court has to be sure 'beyond all reasonable doubt' before it can return a guilty verdict. The court therefore requires a level of certainty that science can rarely provide. Indeed, science is based upon hypotheses and a scientific hypothesis is one that can be proved wrong – provided that one can find the evidence. Organisms are acted upon by numerous internal and external factors and therefore the evidence based upon them usually has to have qualifications attached to it. For example, suppose that the pollen profile found on mud attached to the suspect's shoes matched that found at the site of the crime: this suggests a possible association but it would be impossible to state beyond reasonable doubt that there are not other sites that might have similar profiles – unlikely perhaps, but not beyond doubt. Lawyers are, quite correctly, experts at exploiting the potential weaknesses of biological evidence because it is seldom possible for one to state that there is no alternative explanation for the findings or an event would never happen. Within civil courts, biological evidence has greater potential because here the 'burden of proof' is based upon 'the balance of probabilities'.

This book is intended for undergraduates who may have a limited background in biology and not for the practicing forensic scientist. I have therefore attempted to keep the terminology simple whilst still explaining how an understanding of biological characteristics can be used to provide evidence. With such a large subject base, it is impossible to cover all topics in depth and readers wishing to identify a maggot or undertake polymerase chain reaction analysis should consult one of the more advanced specialist texts in the appropriate area. Similarly, those wishing more detailed coverage of individual cases would be advised to consult the excellent books by Erzinclioglu (2000), Goff (2000), Greenberg and Kunich (2002) and Smith (1986). Where information would not otherwise be easily accessible to undergraduate students, I have made extensive use of web-based material although the usual caveats apply to such sources.

At the start of each chapter, I have produced a series of 'objectives' to illustrate the material covered. They are written in the style of examination essay questions, so that the reader might use them as part of a self-assessment revision exercise. Similarly, at the end of each chapter I have produced a number of short-answer questions to test knowledge and recall of factual information. Also at the end of each chapter, I have made some suggestions for undergraduate projects. Because the usefulness of biological material as forensic evidence depends on a thorough understanding of basic biological processes and the factors that affect them, there is plenty of scope for simple projects based upon identifying species composition or measuring growth rates. Obviously, for the majority of student projects, cost, time and facilities will be serious constraints; DNA analysis can be extremely expensive and requires specialist equipment. Similarly, the opportunities to work with human tissues or suitably sized pigs may not exist. However, worthwhile work can still be done using the bodies of laboratory rats and mice or meat and bones bought from a butcher as substitute corpses, with plants and invertebrates as sources of evidence.

1 The decay process

Chapter outline

The stages of decomposition: fresh; bloat; putrefaction; putrid dry remains

Factors affecting the speed of decay: burial underground; burial underwater; time of year and temperature; exposure to sunlight; burning; geographical location

Objectives

Compare the chemical and physical characteristics of the different stages of decomposition.

Explain how a body's rate of decomposition is affected by the way in which death occurred and the environment in which it is placed.

Compare the conditions that promote the formation of adipocere and of mummification and how these processes preserve body tissues.

The stages of decomposition

After a person dies, their body undergoes dramatic changes in its chemical and physical composition. These changes can provide an indication of how long a person has been dead (Dix, 2000) but they also influence the body's attractiveness to detritivores – organisms that consume dead organic matter. An understanding of the decay process and the factors that influence it is therefore helpful in the study of animals, plants and microbes associated with dead bodies. The stages of decomposition in terrestrial environments can be loosely divided into four stages: fresh, bloat, putrefaction and putrid dry remains. However, these stages merge into one another and it is impossible to separate them into discrete entities.

There are a number of specialist terms that are used in association with the study of human remains. If there is a desire to know more about the causes of

Essential Forensic Biology, Alan Gunn
© 2006 John Wiley & Sons, Ltd

a person's death then the body is subjected to an 'autopsy' or 'postmortem examination', although in the majority of deaths there are no suspicious circumstances so no autopsy takes place. Some doctors have expressed concern about this because it is estimated that 20–30% of death certificates incorrectly state the cause of death (Davies *et al.*, 2004). However, all bodies discovered in suspicious circumstances are subjected to an autopsy and doctors who have received advanced training in pathology perform these. Pathology is the study of changes to tissues and organs caused by disease, trauma and toxins, etc. and forensic pathologists specialize in cases in which death has occurred in suspicious circumstances. The study of what happens to human remains after a person dies is known as 'taphonomy' and the factors that affect the remains are known as 'taphonomic processes'. Thus, temperature, maggot feeding and cannibalistic practices are all examples of taphonomic processes and they are of interest to both forensic pathologists and to archaeologists uncovering the graves of persons who died many hundreds of years ago.

When investigating any death it is essential to keep an open mind as to the possible causes. For example, if the partially clothed body of a woman is found on an isolated moor, there are many explanations other than she was murdered following a sexual assault. First of all, she may have lost some of her clothes after death through them decaying and blowing away or from them being ripped off by scavengers (see Chapter 8). Secondly, she may have been a keen rambler who liked the open countryside. Most people die of natural causes and she may have suffered from a medical condition that predisposed her to a heart attack, stroke or similar potentially fatal condition whilst out on one of her walks. Another possibility is that she may have committed suicide: persons with suicidal intent will sometimes choose an isolated spot in which to die. For example, when Dr Richard Stevens, a consultant haematologist at the Royal Manchester Children's Hospital in Pendlebury, disappeared from work in July 2003 he sparked a nationwide hunt that lasted for many months. There were several false sightings but ultimately his body was discovered by accident many miles away at the back of a walk-in mineshaft on a remote mountainside in Cumbria. Here, he had committed suicide by injecting himself with a combination of two drugs that resulted in his death. Another explanation for the woman's death would be that she had suffered an accident, such as tripping over a stone, landing badly and receiving a fatal blow to her head. And, finally, it is possible that she was murdered. All of these scenarios must be considered in the light of the evidence provided by the scene and the body.

Fresh

When someone's heart stops pumping blood around their body, the tissues and cells are deprived of oxygen and rapidly begin to die. Different cells die at different rates, so, for example, brain cells die within 3–7 minutes while skin cells can be taken from a dead body for up to 24 hours after death and still grow in a laboratory culture. Contrary to folklore, human hair and fingernails do not continue to grow after death, although shrinkage of the skin can make it seem as though they do. It is important to bear in mind that as the joints and muscles relax following death, a person's height may increase by as much as 3 cm.

Once cells start to die and decay their DNA begins to be degraded into ever-smaller fragments: this presents both potential opportunities and problems for the forensic scientist. The opportunity comes from the observation by DiNunno et al. (2002) that within liver cells there is an almost linear correlation between the degree of degradation of the DNA and the time since death. This would therefore suggest a means of estimating how long a person had been dead, although more work is required to confirm how susceptible the technique is to interference by environmental factors and underlying medical conditions such as liver disease. An accurate knowledge of the time since death is often crucial to a homicide or suspicious death investigation but all current methods for its estimation have their drawbacks. It is therefore always helpful if at least two or three different methods are employed so as to increase confidence in the results. The problems of DNA degradation are much more important because once this occurs its usefulness as an identifier of either the victim or the culprit of a crime can be compromised. DNA degrades more quickly in some tissues than in others and as a rule it is best extracted from the bones of the femur or ribs or from the molar teeth.

Between 20 and 120 minutes after death, *livor mortis* (also called hypostasis, and postmortem lividity) is usually seen – it can be found in all bodies but may be difficult to observe. *Livor mortis* is a purple or reddish purple discoloration of the skin caused by the blood settling in the veins and capillaries of the dependent parts of the body. It starts as a series of blotches that then spread and deepen in colour with time. Initially the blood remains in the blood vessels but with time the blood cells haemolyse (break down and rupture) and the pigment diffuses out into the surrounding tissues, where it may be metabolized to sulphaemaglobin that gives rise to a greenish discoloration. Sulphaemoglobin is not present in normal blood although it may be formed after exposure to drugs such as sulphonamides. This emphasizes the need to be aware that normal decomposition processes may mimic those that are induced before death or by the action that induced death.

Blood remains liquid within the circulatory system after death, rather than coagulating, because of the release of fibrinolysins from the capillary walls. These chemicals destroy fibrinogen and therefore prevent clots from forming. However, wounds that are caused after death do not bleed profusely because the heart is no longer beating and blood pressure is not maintained. Blood from even a deeply incised wound therefore trickles out as a consequence of gravity rather than being spurted out, as it might if inflicted during life. In the past, a suspected murderer would sometimes be compelled to touch the wounds on the body of his victim: if the wounds bled, then he was considered guilty. Clearly, because the blood does not clot after death, it would be difficult for a person to touch a wound without staining his hands. Unlike the situation on land, in the case of drowning or a dead body disposed of in a lake or river there may be a considerable loss of blood from wounds. After initially sinking, a dead body tends to rise to the surface, owing to the accumulation of gas from the decay process, and then floats face downwards. Consequently, the blood pools in the facial and dorsal regions and wounds affecting these areas after death may bleed profusely.

After about 10–12 hours of a body remaining in a set position, the discoloration caused by *livor mortis* becomes fixed (Figure 1.1). If the body is then moved and left in a different position a second area of discoloration forms. Two or more distinct patterns of discoloration therefore indicate movement of the body. Pressure, whether from tight-fitting clothes such as belts and bra straps, a ligature around the neck, ropes used to bind hands together or corrugations in the surface on which the body is resting, will prevent the underlying blood vessels from filling with blood and therefore these regions will appear paler than their surroundings – this is known as 'pressure pallor' or 'contact pallor'. Whilst the body is fresh, it is possible to distinguish between *livor mortis* and bruising because in the former the blood is restricted to the dilated blood vessels whilst a bruise results from the leaking of blood into the surrounding tissues, and the formation of clots, during life. However, as the body decays it becomes more difficult to distinguish between the two.

The rate of development of *livor mortis* varies from body to body and is also influenced by underlying medical conditions, such as circulatory disease. Consequently, there is some variation in the literature about when events begin, when they reach their maximal effect and when *livor mortis* becomes 'fixed' (DiMaio and DiMaio, 2001; http://www.dundee.ac.uk/forensicmedicine/lib/

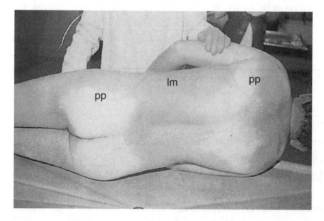

Figure 1.1 Characteristic pattern of *livor mortis* (lm) and pressure pallor (pp) resulting from a dead body lying on its back. The reddening results from the settling of blood in the veins whilst the pale regions are where the pressure of the body against the underlying substrate has constricted the vessels. (Reproduced from Shepherd, 2003, with permission from Arnold)

timedeath.htm). Therefore, on its own, *livor mortis* has limited usefulness in determining the time since death.

Approximately 3–4 hours after death, *rigor mortis* (the stiffening of muscles and limbs) sets in and the whole body becomes rigid by about 12 hours. The condition can, however, be broken by pulling forcefully on the affected limbs. Rigor commences with the small muscles and those being used most actively prior to death. It is brought about by the rise in the intracellular concentration of calcium ions in muscle cells that follows death, as the membranes around the sarcoplasmic reticulum and the cell surface become leaky and calcium ions are therefore able to move down their concentration gradient into the sarcoplasm. This rise causes the regulatory proteins troponin and tropomyosin to move aside, thereby permitting the muscle filaments actin and myosin to bind together and form cross-bridges. This is possible because the head of the myosin molecule would already be charged with ATP before death. However, actin and myosin once bound, are unable to detach from one another because this process requires the presence of ATP – and this is no longer being formed. Thus, the actin and myosin filaments remain linked together by the immobilized cross-bridges, resulting in the stiffened condition of dead muscles. Subsequently, *rigor mortis* gradually subsides as the proteins begin to degrade and it disappears after about 36 hours. *Rigor mortis* is prolonged at low temperatures and at a constant 4°C may last for at least 16 days with partial stiffening still detectable up until 28 days after death (Varetto and Curto, 2004). The extent and degree of *rigor mortis* is therefore not an especially accurate measure of time since death.

Unlike *rigor mortis*, 'cadaveric rigidity' sets in immediately after death. It is rare and is said to be associated with individuals who were extremely stressed, emotionally and physically, immediately before they died (Shepherd, 2003). One would have thought that this would include most murder victims and also many who die of painful medical conditions, so there must be other factors, possibly genetic, to explain the rarity. The physiological basis of this form of rigidity is not known.

Bloat

The intestines are packed with bacteria and these do not die with the person. These micro-organisms start to break down the dead cells of the intestines, while some, especially the *Clostridia* and the *coliforms*, start to invade the other body parts. At the same time, the body undergoes its own intrinsic breakdown, known as autolysis, that results from the release of enzymes from the lysosomes (subcellular organelles that contain digestive enzymes), thereby causing cells to digest themselves and chemicals, such as the stomach acids, from the dead cells and tissues. The pancreas, for example, is packed with digestive enzymes, and so rapidly digests itself. Autolysis may also occur on a more restricted scale in a living person as a consequence of certain pathological processes.

The decomposing tissues release green substances and gas that make the skin discoloured and blistered, starting on the abdomen in the area above the caecum (Figure 1.2). The front of the body swells, the tongue may protrude and fluid from the lungs oozes out of the mouth and nostrils. This is accompanied by a terrible smell because gases such as hydrogen sulphide and mercaptans (sulphur-containing organic molecules) are produced as end products of bacterial metabolism. Methane (which does not smell) is also produced in large quantities and contributes to the swelling of the body. In the UK, this stage is reached after about 4–6 days during spring and summer but would take longer during colder winter weather.

Blowflies and other detritivores are attracted by the odour of decomposition (Figure 1.3), and as the smell changes during the decomposition process so does the species of invertebrate that is attracted. Therefore, species that are attracted to 'fresh corpses' are often different to those that are attracted to corpses in an advanced state of decay (Table 1.1). Blowflies do not lay their eggs on corpses once these have passed a certain state of decomposition or they have become dry or mummified. By contrast, dermestid beetles do not colonize corpses until these have started to dry out (for more details see Chapters 6 and 7).

Putrefaction

Some authors distinguish several stages of putrefaction (decay) but the usefulness of this is uncertain. Different parts of the same body may decay at different rates and decay itself is a gradual process rather than a series of discrete

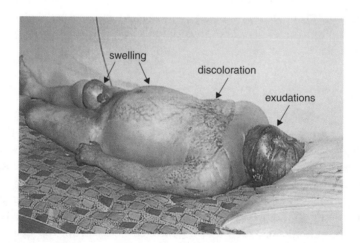

Figure 1.2 Late bloat stage of decomposition. The body is about 7 days old and exhibits pronounced swelling owing to accumulation of gas. Note discoloration of the skin and exudates from the mouth and nose. (Reproduced from Shepherd, 2003, with permission from Arnold)

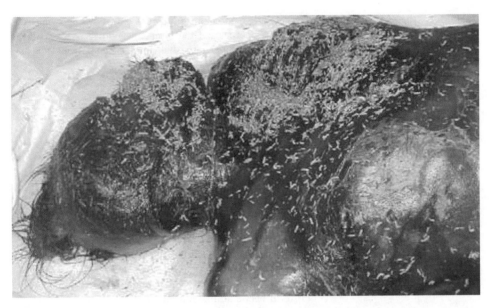

Figure 1.3 Blowfly maggots developing upon a human corpse. Note how mature maggots can be seen crawling over the surface and the discoloration of the skin. (Reproduced from Klotzbach *et al.*, 2004, with permission from Elsevier)

events. As the body enters the bloat stage, it can be said to be 'actively decaying' and during this time the soft body parts rapidly disappear as a result of autolysis and microbial, insect and other animal activity. The body then collapses in on itself as gases are no longer retained by the skin. At this point, the body is said to enter a stage of 'advanced decay' and, unless the body has been mummified, much of the skin will now be lost. Obese people tend to decay faster than those of average weight and this is said to be due to the 'greater amount of liquid in the tissues whose succulence favours the development and dissemination of bacteria' (Campobasso *et al.*, 2001). At first sight, this appears surprising because fat has a lower water content than other body tissues and obese individuals therefore have a lower than normal water content. However, fat can act as a 'waterproofing', preventing the evaporation of water and therefore the drying out of the corpse whilst its metabolism yields large amounts of water.

Adipocere (grave wax) may be formed during the decay process if the conditions are suitable and this is capable of influencing the future course of decay (Fiedler and Graw, 2003; Forbes *et al.*, 2004). It is a fatty substance that is variously described as being whitish, greyish or yellowish and with a consistency ranging from paste-like to crumbly. Once formed, it acts to inhibit further decomposition and the body can remain virtually uncorrupted for many years (Figure 1.4). Adipocere formation is therefore a nuisance in municipal

Table 1.1 The sequence in which insects arrive to colonize a corpse during the decomposition process. The stages of decay merge into one another and the insects may arrive or leave sooner or later than is indicated in the table, depending upon the individual circumstances (for more details see Chapters 6 and 7)

Stage of decomposition	Insect
Fresh	Blowfly eggs and 1st instar larvae Fleshfly 1st instar larvae Burying beetle adults
Bloat	Blowfly eggs + 1st, 2nd, 3rd instar larvae Fleshfly 1st, 2nd, 3rd instar larvae Burying beetle adults and larvae Histerid beetle adults and larvae
Putrefaction	No blowfly eggs once advanced putrefaction Blowfly 2nd, 3rd instar larvae Fleshfly 2nd, 3rd instar larvae Blowfly and fleshfly larvae leaving corpse for pupation site Histerid beetle adults and larvae Eristalid fly larvae (liquified regions) Phorid fly larvae (later stages of putrefaction) Piophilid fly larvae (later stages of putrefaction)
Putrid dry remains	No blowfly larvae Stratiomyid fly larvae Dermestid beetle adults and larvae Tineid moth larvae Pyralid moth larvae

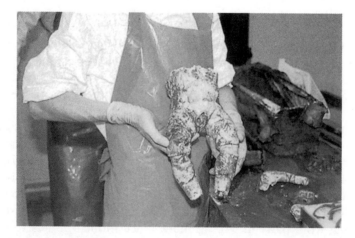

Figure 1.4 The formation of adipocere has preserved the body of this child despite it being buried for about 3 years. (Reproduced from Shepherd, 2003, with permission from Arnold)

graveyards because it prevents the authorities from recycling grave plots but it is very useful to forensic scientists and archaeologists who wish to autopsy long-dead bodies.

The term adipocere refers to a complex of chemicals rather than a single chemical compound and it results from the breakdown of body lipids. As soon as a person dies, autolysis and bacterial decomposition of triglycerides, which make up the majority of the body's lipid stores, results in the production of glycerol and free fatty acids. The free fatty acids are comprised of a mixture of both saturated and unsaturated forms, but as adipocere formation progresses the saturated forms become predominant. The formation of these fatty acids lowers the surrounding pH and thereby reduces microbial activity and further decomposition. The formation of adipocere also brings about a characteristic odour, the nature of which changes with time and has been used to train cadaver dogs to detect dead bodies. Extensive adipocere formation can result in the body swelling and consequently the pattern of clothing, binding ropes or ligatures can become imprinted on the body surface whilst incised or puncture wounds may be closed and difficult to detect. Adipocere formation is not exclusive to human decomposition (Forbes et al., 2005d) and this should be borne in mind if there is a possibility that human and animal remains have become mixed together. For example, the bodies of animals are often found at the bottom of disused mine shafts, having stumbled in or been thrown in by a farmer looking for a quick means of disposing of dead livestock. Murderers will also make use of such facilities.

Adipocere formation has been described from bodies recovered from a wide variety of conditions, including fresh water, seawater and peatbogs, shallow and deep graves or tightly sealed containers, and in bodies buried but not enclosed at all (e.g. Evershed, 1992; Mellen et al., 1993). Some authors mention that warm conditions may speed its formation but adipocere has been recorded from bodies recovered from seawater at a temperature of 10–12°C and from icy glaciers (Ambach et al., 1992; Kahana et al., 1999) – the preservation of the 5300-year-old 'Iceman' found in the Tyrol region appears to be at least partly a consequence of the formation of adipocere (Bereuter et al., 1997; Sharp, 1997). A wide variety of times have also been cited for the time taken for adipocere formation to become extensive, ranging from weeks to months to over a year. Obviously, the time will be heavily dependent upon the local conditions and it is not yet possible to use the formation of adipocere as an estimate of the time since death. However, adipocere does leak out of the body and into the surrounding soil and therefore its presence might be useful in determining if a body had been stored in a particular location but then moved or if the extent of adipocere formation in the body matches that which might be expected in the surrounding soil if the body had been there since death. Forbes et al. (2005a, b, c) conducted an extensive series of experiments on the physical and chemical factors promoting the formation of adipocere. They found that adipocere would form in soil types ranging from sandy to clayey, provided that the soils were kept moist, and also in sterile soil that had been heated at 200°C for 12 hours to remove the normal soil microbial flora. 'Bodies' buried directly in the ground tended to form adipocere more rapidly than those contained in a coffin. Interestingly, placing the 'body' in a plastic bag retarded the formation of adipocere but if the 'body' was clothed and then placed in the plastic bag adipocere formation was promoted. They suggested that this was owing to the clothing absorbing glycerol and other decay products that would otherwise inhibit the pathways through which adipocere is formed. Polyester clothing was deemed to be the most effective, probably as a consequence of its ability to retain water and, compared to cotton clothing, resistance to decay.

Another means by which a body might become naturally preserved is through mummification. This occurs when it is exposed to dry conditions coupled with extreme heat or cold, especially if there is also a strong air current to encourage the evaporation of water. It is typically seen in persons who have died in deserts, such as the hot Sahara and the cold Tibetan plateau. It is also found in murder victims who have been bricked up in chimneys or persons who have died in well-sealed centrally heated rooms. Size is important, and dead babies, owing to their large surface area to volume ratio, will lose water more rapidly than an adult. Newly born babies lack an active gut microbial flora and therefore not only will they lose water quickly but they also may dehydrate before microbial decomposition can cause major destruction of tissues.

Putrid dry remains

After the skin and soft tissues are removed, the body is reduced to the hard skeleton and those structures that are more difficult to break down, such as the tendons, ligaments, fingernails and hair. Organs such as the uterus and prostate glands are also fairly resistant to decay and may last for several months if the body is kept in a well-sealed container. Because there are still traces of dead organic matter being broken down by bacteria and fungi, a skeletonized body still smells of decay.

Once the skin and the soft tissues have been lost, it can be difficult to detect death that results from penetrating wounds, such as those caused by stabbing with a knife or screwdriver. However, in about half of such cases there is usually associated damage to the bones and cartilage (Banasr et al., 2003) so it is important that these tissues are examined very carefully.

For a summary of the decomposition stages and their characteristics, see Table 1.2.

Factors affecting the speed of decay

The speed with which a body decays depends upon many factors (see Table 1.3).

Table 1.2 Summary of stages of decomposition and their characteristics

Stage of decomposition	Characteristics
Fresh	Body starts to cool and autolysis begins. *Livor mortis* and *rigor mortis* may be seen
Bloat	Discoloration of skin surface, body swells from accumulation of gases. Tongue protrudes, fluid expelled from orifices. Soft tissues visibly decaying. Rapid decay owing to intense microbial and invertebrate activity
Putrefaction	Progressive loss of skin and soft tissues. Body deflates as decomposition gases escape. Decay owing to invertebrate and microbial activity starts to slow down once soft tissues removed and body starts to dry out
Putrid dry remains	Skin and soft tissues lost. Decay proceeds more slowly. Progressive loss of uterus/prostate gland, tendons, cartilage, fingernails, hair. Skeleton may become disarticulated through environmental and biological processes

Burial underground

Some species of blowflies and fleshflies will lay their eggs on a corpse within minutes of a person dying and will travel long distances to do so. The rapidity and extent to which a corpse is colonized by the larvae of these flies, along with the activities of other invertebrates, microbes and vertebrates such as dogs and birds, heavily influences the speed with which a body decays. Consequently, those factors that restrict their access or reduce their activity, such as lack of oxygen or the temperature being too low or too high, reduce the rate of decomposition enormously. For example, buried corpses decay approximately four times slower than those left on the surface, and the deeper they are buried, the longer they are preserved, provided that the ground is not waterlogged (Dent *et al.*, 2004). Even a shallow covering of soil will exclude blowflies from colonizing a body and hence reduce the rate of decomposition, although some fly species, such as the coffin fly (*Conicera tibialis*), are able to locate bodies a metre or more below ground. Sealing a body within an airtight container also reduces its rate of decay because of the reduced oxygen level and the inability of invertebrates to gain access to it. Bodies that are encased in concrete will also be preserved, although there are few experimental data on how the body's chemistry changes with time under these conditions. A body that is left hanging may decay more slowly than one that is lying on the surface of the ground (Wyss and Cherix, 2004). This is probably because when a body is suspended in mid air there would not be a moist, dark, under-surface where flies could lay their eggs,

Table 1.3 Summary of factors promoting or delaying the rate of decay

Factors promoting decay	Factors delaying decay
Oxygen supply not restricted	Oxygen supply restricted
Warm temperature (15–37°C)	Cold temperatures (<10°C; decay will cease below 0°C)
Humid atmosphere	Dry atmosphere
Presence of invertebrate detritivores (e.g. blowfly larvae)	Absence of invertebrate detritivores
Wasp, ant and other invertebrate predators feeding on corpse	Wasp, ant and other invertebrate predators feeding on detritivores
Wounds permitting invertebrates easier access to internal body tissues	Inability of detritivores to gain access to all or part of the corpse
Surface burning causing skin to crack and thereby allowing easier access of invertebrates and oxygen to internal tissues	Intense burning resulting in tissues carbonized and drying out.
Obesity	
Suffering from septicaemia or myiasis before death	
Body exposed to the environment above ground	Burial on land or underwater (rate of decay declines with increasing depth)
Body resting on soil	Body suspended above ground (e.g. hanging)
	Formation of adipocere
	Mummification

the circulation of air would promote drying out and many maggots would fall off whilst crawling around or be washed off by the rain.

Burial underwater

Decomposition underwater is typically twice as slow as when the body is exposed to air and the underwater rate is even slower at low temperatures and oxygen levels. Under deep-sea conditions, decomposition may be prolonged and fish, crabs, starfish and other benthic (i.e. those living on or close to the sea bed) invertebrates may be more important than microbial decomposition. Bodies that float on the surface of ponds, lakes and rivers can still be colonized by blowfly larvae and this, along with the exposure to higher temperature and oxygen levels, increases the rate of decomposition.

Time of year and temperature

Many invertebrates show distinct patterns of seasonal activity. For example, in the UK, many blowfly species are not active during the winter months and consequently a body tends to decay more slowly during this time. Temperatures that are too high or too low will also restrict the activity of invertebrates and microbes and thereby reduce the rate of decay.

Burning

Murderers often attempt to dispose of their victim's body by burning the corpse. However, they are seldom completely successful owing to the extremely high temperatures required – identifiable human remains may still be found among the ashes produced by a crematorium, which typically operates at over 1000°C. Badly burnt bodies are also a common feature of victims of explosions and traffic and aircraft accidents. Some factors associated with burning may reduce the rate of decomposition whilst others promote it, so it is difficult to generalize. For example, burning sterilizes the skin surface and dries the underlying tissues, making them unsuitable for the growth of microbes and blowfly maggots but it also causes cracks through which they may invade the deeper tissues that are less affected. Similarly, although the skin surface may be charred, the temperature may not have been high enough to affect the gut microbial flora, so decomposition may commence here as normal. Burning induces chemical changes in proteins, carbohydrates, lipids and other organic molecules that may affect their suitability to support microbial and maggot growth, but there is little published information on this. Some workers have found that burnt corpses retain their attractiveness to blowflies whilst others have found that it reduces their likelihood to lay eggs on the body (Catts and Goff, 1992; Avila and Goff, 1998). Obviously, a great deal depends on the degree of burning and the individual circumstances.

Exposure to sunlight

Many invertebrates avoid laying their eggs on and colonizing regions of a corpse that are exposed to the full summer sun. This is because it would cause their delicate eggs and young larvae to be killed by a combination of desiccation and exposure to UV light. However, eggs may still be laid on the under-surfaces and if the body is clothed or similarly covered then the covered regions will be protected from the sun and have a more humid microenvironment. This can provide a more suitable environment for microbes and maggots to grow and consequently they may decay more rapidly than exposed body parts. Loosely fitting clothing will facilitate the entry of invertebrates and the circulation of air and therefore enhance the rate of decay, whereas tight-fitting clothing may delay it.

Geographical location

The abundance and species composition of the invertebrate fauna vary considerably between regions and this can affect the speed with which a body is located, colonized and decomposed. For example, because cities offer warm microclimates, some blowfly species may remain active throughout the winter period whereas they would be inactive in the surrounding countryside. In both cities and the countryside, the adults of some blowfly species enter buildings during the autumn period and will attempt to overwinter indoors. Should a body be placed within a building where they occur and the temperature is high enough for them to be active, then colonization of the corpse will commence even in the depths of winter. Decay proceeds much faster in the tropics, where conditions are both hot and humid, and slower in cold or dry conditions. In the tropics, a corpse can become a moving mass of maggots within 24 hours but in the UK it would take several days to reach this stage, even during the summer.

Quick quiz

1. Distinguish between *livor mortis* and *rigor mortis*.

2. What is meant by the term 'taphonomic process'? State two examples of taphonomic processes.

3. What causes the 'bloat' stage of decomposition?

4. What is adipocere and what is its forensic relevance?

5. Why do newly born babies mummify more readily than older children and adults?

6. Why does a buried body decay more slowly than one lying on the surface of the ground?

7. Why does a hanging body decay more slowly than one lying on the surface of the ground?

8. During winter, why might an exposed body decay faster if left in the centre of a city than in the outlying countryside?

9. State one means by which burning can reduce the rate of decay of a corpse and one means by which its rate of decay might be increased.

Project work

Title

The effect of freezing on the rate of decay.

Rationale

Murderers sometimes store their victim's body in a freezer before disposing of it. Freezing will cause tissue damage so once the body has thawed does it decay at the same rate as an unfrozen body?

Method

Bodies or tissues can be frozen for varying lengths of time and then placed above or below ground and the rate of decay, speed of colonization by invertebrates, etc. can be compared to a control unfrozen body. If a thermocouple is placed on the frozen body when it is exposed, it would be possible to determine whether blowflies are deterred from laying until the surface temperature has risen to near-ambient levels. Histological changes could also be assessed, along with biochemical assays to determine the speed with which autolysis begins.

Title

The effect of burying in concrete on the rate of decay.

Rationale

Bodies are sometimes disposed of in the concrete foundations of buildings or bridges.

Method

Bodies or tissues, which may or may not be wrapped in clothes, would be encased in concrete and then left at varying temperatures. After varying times the body would be retrieved from the concrete and its state of decay compared to control bodies that were not placed in concrete. The ability to extract DNA from the bodies would be assessed and structural changes to the surrounding concrete determined.

2 Body fluids and molecular biological techniques as forensic indicators

Chapter outline

Body fluids and waste products
    Blood cells and blood typing
    Confirming the presence of blood
    Bloodstain pattern analysis: the Sam Sheppard case; the Sion Jenkins case
    Saliva and semen as forensic indicators
    Faeces and urine as forensic indicators
Molecular biology
    The structure of DNA
    DNA profiling: the Colin Pitchfork case
    Forensic applications of DNA profiling
    Evaluation of DNA evidence
    Polymerase chain reaction
    Short tandem repeat markers
    Y-Short tandem repeat markers
    Single nucleotide polymorphism markers
    Mobile element insertion polymorphisms
    Mitochondrial DNA: the Tsar Nicholas II case

Objectives

Describe how the deposition and composition of blood at a crime scene can be used to determine how the crime was committed and identify both the victim and the assailant.

Explain how semen, urine and faeces found at a crime scene can be used as forensic indicators.

Discuss what is meant by the term 'DNA profiling' and evaluate the different methods currently available by which it can be conducted.

Evaluate the advantages and limitations of the different DNA-based methods for determining gender.

Essential Forensic Biology, Alan Gunn
© 2006 John Wiley & Sons, Ltd

Body fluids and waste products

Blood cells and blood typing

Blood consists of a variety of different cell types, collectively known as blood cells, which are suspended in a watery fluid called serum. Traditionally, the study of blood is referred to as haematology whilst the study of serum (and in particular the immune factors within it) is called serology. Typically, the intrusion of forensic science has complicated things and the term 'forensic serology' is sometimes taken to encompass not only the study of serum but also blood cells, saliva and semen for forensic purposes.

Mature human blood cells can be divided into those that posses a nucleus and those that do not: both types can provide forensic information. Those lacking a nucleus are the red blood cells (RBC) (also known as erythrocytes) and the platelets. The red blood cells possess the pigment molecule haemoglobin that gives blood its red colour and is responsible for the transport of oxygen and carbon dioxide. The platelets are smaller than the red blood cells and, unlike them, they are capable of amoeboid-like movement, i.e. they move by sending out cell processes in a similar manner to the single-celled organism 'amoeba' and consequently often look star-shaped when viewed using a microscope. Platelets are responsible for the clotting mechanism and the production of chemical growth factors that maintain the integrity of the blood vessels. The cells that contain nuclei are collectively known as white blood cells (WBC) or leucocytes. There are many different sorts of white blood cell but all of them are capable of amoeboid-like movement and they are responsible for the body's immune defence capabilities. All the white blood cells contain nuclei and they also have mitochondria – subcellular organelles that produce energy for the cell – therefore, unlike the red blood cells, they posses DNA.

All cells within the body carry upon their outer surface an array of molecules called 'antigens'. These antigens act like a passport and identify the cell as a legitimate 'citizen of the body'. The antigens are recognized by a group of proteins, called 'antibodies', that are secreted by certain white blood cells and serve as 'immigration control'. If a cell does not posses the correct antigens on its surface, it is deemed to be a foreigner and an immune response is mounted to destroy it. The nature of the antigens found on the surface of red blood cells is of medical importance because unless the correct blood type is used during a blood transfusion it would be rejected by the recipient's body and thereby result in fatal consequences. The most common antigens found on red blood cells are those that comprise the ABO system. The ABO system works as follows: an individual's red blood cells may posses either only class A antigens (type A), only class B antigens (type B), both classes A and B antigens (type AB) or neither class A nor class B antigens (type O). Persons who are type A tolerate their own class A antigens but produce antibodies against B antigens that bind to the surface of the red blood cells and form bridges between them. This causes the

Table 2.1 Summary of ABO blood group interactions (Ag = antigens, Ab = antibodies, RBC = red blood cells)

Characteristic	Blood type			
	A	B	AB	O
Ag on RBC	A	B	Both A and B	Neither A nor B
Ab in plasma	Anti-B	Anti-A	Neither anti-A nor anti-B	Both anti-A and anti-B

red blood cells to agglutinate (clump together) – a reaction that occurs quickly and can be observed using a microscope. In a similar manner, type B persons produce antibodies against class A antigens, type AB persons produce antibodies against neither class A nor class B antigens and type O persons produce antibodies against both class A and class B antigens. These interactions are summarized in Table 2.1.

There are many other groups of antigens that are found on red blood cells, the most well known being those responsible for the Rhesus (Rh) factor (also known as antigen-D and agglutinogen-D). The Rhesus factor is so called because the antigens responsible for it were first described in Rhesus monkeys. Persons who posses these antigens are said to be Rhesus positive [Rh(+)] whilst those who do not are Rhesus negative [Rh(–)]. Eighty-three per cent of the UK population are Rh(+) and most belong to either blood types O (44 per cent) or A (42 per cent). Although we talk of a person being blood types A, B, AB or O and being Rh(+) or Rh(–), in actual fact all of these characteristics can themselves be divided up into many more sub-combinations (e.g. O1 and O2). Certain races tend to have higher proportions of particular blood groups than others – for example type AB is more common among the Japanese (10 per cent) than among Europeans (UK, 4 per cent). More precise details of racial origin may be obtained from finding evidence of rare inherited disease traits or particular antigens. For example, sickle cell anaemia occurs almost exclusively among Black people of African descent. Similarly, there is a variety of the Duffy antigen (called phenotype a-b-) that is extremely common among West Africans and their descendents but almost absent from White and Asian populations. Both of these variations are considered to be protective against malaria because the *Plasmodium* parasite finds it harder to recognize and invade red blood cells expressing these traits. By contrast, the Kell antigen is predominantly found among White persons (Reid and Lomas-Francis, 1996). In addition to blood type and the Rhesus factor, many of the enzymes and other proteins found in red blood cells are also polymorphic, i.e. they exist in more than one form. Consequently, even though an individual may have a common blood type such as type O, once all the other variables are taken into account his combination may be shared by only a small number of the population. If a total of eight serological variables are used, it is generally estimated that the chances of two unrelated persons sharing the same profile is between 0.01 and 0.001 (i.e. between 1 in 100 and 1 in a 1000). This estimate is known as the match probability (Pm). This means that blood typing can provide a means of identification, although it is not as accurate as DNA profiling in which match probabilities of up to 10^{-13} have been claimed. Matching a person's blood profile to traces found at a crime scene might be incriminating but on its own would be insufficient to prove guilt. It is, however, effective at excluding suspects and therefore allowing the police to concentrate their resources on more profitable lines of enquiry. Blood typing suffers from several further drawbacks as a forensic tool. Firstly, by comparison with DNA profiling, it requires relatively large amounts of sample and is therefore of limited use where only small specks of blood are available. Serological markers degrade quickly and consequently the amount of information that can be obtained from either a body or a bloodstain that is several days old is reduced. The procedure is further compromised by the interference caused by enzymes released by contaminative bacteria. Interference can also be a problem where the bloodstain contains both the assailant's and victim's blood (e.g. after a fight), if one of the parties has recently received a blood transfusion (in which case their blood profile may be temporarily altered) or following a rape in which semen is left inside the victim's body and therefore will be diluted by his or her serological profile. In addition, the investigation may require a suspect to provide a blood sample. This is an invasive medical procedure with which he or she may legitimately refuse to cooperate.

Confirming the presence of blood

Bloodstains are not always obvious because of the manner in which they were produced or because the assailant cleaned up the crime scene after committing an assault. However, once blood is spilt, it can be extremely difficult to remove all the traces. Any attempt at cleaning up inevitably means using either the kitchen or bathroom as a 'base' – so it is a good place to start looking for blood. In the cleaning process, blood can flow beneath tiles or linoleum or between the boards of wooden flooring. Similarly, if a water tap is grasped with a bloody hand, the blood may flow into the screw mechanism. Consequently, blood can be detected in even an apparently spotlessly clean room if one knows where to look. To highlight the presence of blood an investigator sprays an indicator substance, such as luminol, onto the suspect area. Luminol interacts with the iron found in haemoglobin to produce a faint bluish glow that is best seen in dim light. Unfortunately, luminol interacts with other iron-containing substances and also with copper compounds. Consequently, whilst luminol is highly effective, and can sometimes detect blood on clothing even after it has been washed, a positive reaction only suggests the presence of blood and is not proof. Tests such as this are therefore called 'presumptive tests'.

Presumptive tests are an essential first step in a forensic investigation because many substances can produce stains that give the appearance of blood. For example, there is said to have been a case in which a child was reportedly raped in her own bedroom but once the 'bloodstains' were analysed they turned out to be caused by plum juice. The child apparently took a slice of plum tart from the kitchen and ate it without permission in her bedroom. Presumably, clothing or sheets became stained and the child then invented 'an intruder' to cover up the 'crime'.

There are a number of spot tests that can determine whether a suspicious stain is actually blood but the Kastle-Meyer test is the most commonly used. In this test the stain is treated with a mixture of ethanol (to improve the sensitivity of the test), phenolphthalein (a colour indicator) and hydrogen peroxide (an oxidant). The hydrogen peroxide interacts with the haem molecule of haemoglobin and is broken down into water plus highly reactive free oxygen radicals – these radicals then interact with the phenolphthalein resulting in the solution changing from colourless to pink. In order to be sure the test is working, one should always perform a negative control (e.g. a spot of distilled water) and a positive control (e.g. a spot of dried animal blood) immediately before assessing an unknown stain. If this is not done then the results could easily be called into question. The Kastle-Meyer reaction is also subject to false positives as a result of the hydrogen peroxide being broken down by interactions with other chemicals – these can be found in a variety of substances of which, for some unknown reason, potatoes and horseradishes are the most commonly quoted. Consequently, further tests would be required to improve the reliability of the evidence.

Even if blood is detected, it is not immediately certain that it came from a human being. Household pets are just as likely to fight, scratch and otherwise injure themselves as any human and therefore leave their bloodstains on upholstery and clothing. Similarly, when handling raw meat in the kitchen it is not unusual to leave bloodstains upon surfaces and clothing. Furthermore, it is important to hunt for animal blood when investigating crimes such as badger baiting or when a pet was injured or killed during the course of a break-in or homicide. To determine whether blood comes from a human or an animal one usually requires a precipitin test, which involves reacting the blood sample with anti-human antibodies – these are available commercially and are raised in rabbits. The blood samples and the antisera (anti-human antibodies) are placed in wells punched into an agar gel that is spread over a glass dish or slide. The samples move towards one another through the agar by diffusion or the process can be speeded up using an electric current. If a white line – called the precipitin line – is formed at the point at which the two samples meet, this is indicative of an interaction between the antigens in the blood and antibodies in the rabbit antisera and therefore the blood is human. If no precipitin line forms and there is a suspicion that the blood belongs to an animal, then the procedure can be repeated using antibodies raised against the appropriate animal sera.

DNA-based techniques are now being developed to differentiate human and non-human blood and tissue samples. For example, Matsuda *et al.* (2004) describe a highly specific PCR-based protocol (see later for more information on PCR) that utilizes primers for the human mitochondrial cytochrome *b* gene. They state that following the agarose electrophoresis step of the PCR process, human DNA produces a single band whilst blood from other vertebrates fails to produce any bands at all. Molecular methods such as this have the advantages of being simple, extremely specific and require extremely small amounts of sample material but the reagents tend to be expensive.

Bloodstain pattern analysis

Bloodstain pattern analysis (also called blood splatter or blood spatter analysis) is a specialized branch of forensic science in which the investigator determines the shape and distribution of bloodstains. From this information it is possible to determine firstly whether or not a crime was actually committed and, if so, how the conflict developed and how the blows were inflicted. The earliest record of bloodstains being used in court proceedings relates to the trial of the unfortunate Richard Hunne in 1514. He was being held in Lollard's Tower in London on five charges of heresy but before he could be examined he was found hanged in his prison cell. This obviously upset 'the powers that be' because they then charged him with 13 more counts of heresy and suicide. (Suicide remained a crime in England and Wales up until 1961 and was a serious offence because it meant that the victim's body could not be buried in consecrated ground.) However, bloodstains were found in his cell and after reviewing these and medical evidence it was concluded that: 'whereby it appearth plainly to us all, that the neck of Hunne was broken, and the great Plenty of Blood was shed before he was hang'd. Wherefore all we find by God and all our Consciences, that Richard Hunne was murder'd. Also we acquit the said Richard Hunne of his own Death' (Forbes, 1985).

An adult human contains about 5 litres of blood; loss of approximately 30 per cent blood volume (~1.5 litres) through either internal or external bleeding usually results in a person losing consciousness or becoming severely incapacitated whilst the loss of 40 per cent (~2 litres) blood volume can be fatal. Arterial blood (with the exception of that going from the heart to the lungs) is bright red in colour owing to its high oxygen content whereas venous blood is darker in coloration as it is deoxygenated. However, once shed, blood darkens and begins to clot within about 3 minutes and it is impossible to tell whether it originated from an artery or vein. Once a bloodstain has dried there are currently no reliable tests to determine how old it is. It would therefore be an interesting experiment to determine whether a stain's age could be related to the amount of decomposition within the DNA. Arterial blood flows under high pressure and will spurt out over a considerable distance should the vessel be cut. Further-more, it will continue to spurt out in bursts owing to the beating of the heart, thereby producing a characteristic stain pattern. By contrast, when a vein is cut, the blood tends to seep out rather than spurt because of the comparatively lower blood pressure. Varicose veins result from the valves within the vein becoming leaky – this disrupts the normal flow of the blood, allowing it to pool and therefore cause localized swelling. Where the varicose veins are close to the skin surface, they are vulnerable to being knocked and ruptured. Although not common, elderly sufferers of varicose veins have been known to stagger around their home after breaking open a vein, spreading large amounts of blood and giving the impression that they have been physically attacked. Similarly children (and some adults) often suffer from nosebleeds and even a minor blood loss soon gives the appearance of a serious crime scene.

In the initial stages of an investigation, the characteristics of all suspicious stains at a crime scene are noted. This will include the exact position of every stain, along with its size and shape and the nature of the material on which it was formed (clothing, plastic, wood, etc.): these records are made using pho-tography and a written report so that every spot that is to form part of the evi-dence is given an identifying number for future reference. From the distribution and shape of the stains one can determine how the blood was shed and hence how a wound was inflicted. For example, large (4 mm or more in diameter) cir-cular drops of blood on the floor would indicate that the blood was travelling slowly and that the victim was stationary or hardly moving. This is indicative of blood dripping passively from a wound or a weapon. The shape of these 'passive stains' indicates how far it had fallen – blood falling vertically onto the floor from 1–50 cm tends to form circular drops with slightly frayed edges whereas blood falling from a greater height would form a sunburst pattern (Figure 2.1). As a drop of blood falls, it is held together by surface tension and behaves as though it had an elastic skin. Surface tension is caused by cohesive forces between molecules at the surface of a liquid and results in the drop pulling into its smallest possible area. The consequence of this is that a drop of blood will not fall unless it is subjected to a force greater than its surface tension, and once in flight the drop will tend to stay together until it comes into contact with another object. It is important to take into account the nature of the surface onto which blood has impacted as this can influence the stain shape (Figure 2.2). For example, blood fallen onto concrete is far more distorted than if it fell onto glass. Consequently, any conclusions drawn from the bloodstain pattern found at a crime scene should always be verified by experimentation.

A distribution of small spots usually indicates that the blood was being pro-pelled at medium to high velocity, as would occur after being hit over the head by a baseball bat. These are known as projected bloodstains and they can be subdivided into cast-off bloodstains and impact spatter stains (Figure 2.3). Blood that is forcefully expelled from the body or cast from a bloody knife travels in

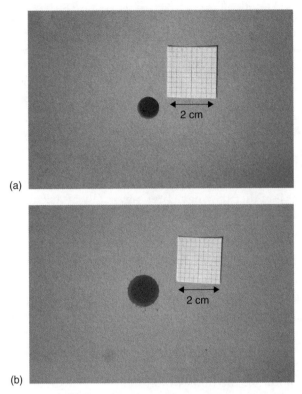

(a)

(b)

Figure 2.1 A blood droplet falling passively from a short distance (in this case 5 cm) onto a smooth hard surface forms a circular-shaped bloodstain (a) whilst the same droplet falling from a greater height (in this case 90 cm) forms a sunburst-shaped bloodstain (b). Square scale = 20 mm

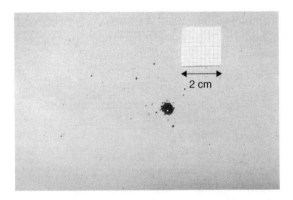

Figure 2.2 The shape of a bloodstain is influenced by the substrate onto which it falls. This bloodstain was formed by a droplet of blood falling passively from 10 cm onto a piece of clean dry calico cloth. The blood droplet disintegrated on impact to produce a patchwork of small surrounding stains. Square scale = 20 mm

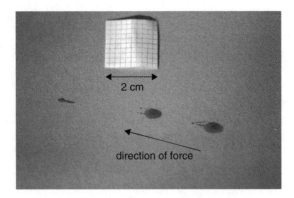

Figure 2.3 Cast-off stains from a bloodstained knife formed on a clean, smooth plastic surface. The direction of the tails indicates the direction of the force

an arc rather than a straight line and where it hits a solid object it produces stains that are elliptical or tear-drop-shaped. The direction of any tails or smaller ends to these stains indicates the direction of the force propelling the blood from the body or weapon. From these shapes it is possible to determine the drops' angles of impact using basic trigonometry. By comparing the angle of impact of different groups of bloodstains it is possible to reconstruct how a weapon was used or a wounded person was moving. Cast-off stains result from blood being thrown from a weapon as it is being wielded. For example, when a knife is pulled outwards and upwards from the body there will be projected a series of bloodstains from the knife blade. The second and any further blows tend to be the ones causing the most blood splatter. Axes and similar hefty weapons are most frequently used with a downward chopping action and the returning upswing leaves a cast-off pattern of blood on the ceiling. Weapons that can pick up a lot of blood may produce cast-off stains on both their outward and return stroke. If there was only one victim and cast-off bloodstains were found in more than one room it indicates that the attack was prolonged and there was a chase. However, if there were cast-off bloodstains in only one room but passive or transfer bloodstains in other rooms it is possible that the attack took place in one room, after which the victim staggered or was dragged elsewhere.

Impact spatter results from impact of a weapon upon the body or upon wet blood. The size of the droplets produced depends upon the force with which the body (or blood) was hit and the tails indicate the direction in which the weapon was being wielded. If lines are drawn through the long axis of a number of groups of these bloodstains, the points at which they intersect, known as the convergence point, will indicate the most likely position of the victim. Even when produced by the same blow, small droplets tend not to travel as far as large droplets because they have less momentum (momentum = mass multiplied by velocity). As the speed of impact increases, the dimensions of the blood droplets decrease and bullets from a powerful gun cause the formation of a mist-like array of tiny stains 1 mm in diameter or less. Even when formed by high-velocity impact, such small droplets of blood seldom travel far. Another way in which small to microscopic droplets of blood may be projected is when blood is coughed or sneezed out of the body, such as after a wound to the nose, mouth or chest. It may also occur when a person is breathing face down through a pool of blood (e.g. after being shot in the head). Bloodstains containing tiny air bubbles are an indicator that the blood was expired.

Transfer bloodstains occur when wet blood is moved between objects and this can result in recognizable prints being left behind (Figure 2.4). For example, a bloody hand will leave prints on doorknobs or weapons whereas a person step-ping in blood will leave behind impressions of his footwear. Similarly, attempts at cleaning up bloodstains will result in smears and smudges, as would the move-ment of a bleeding body across the floor. The pattern of a blood smear can indi-cate the direction in which a person or bloodstained object was dragged: the smear usually begins as a series of drops and these then become ragged along one edge, indicating the direction of travel. The initial spots may be disrupted

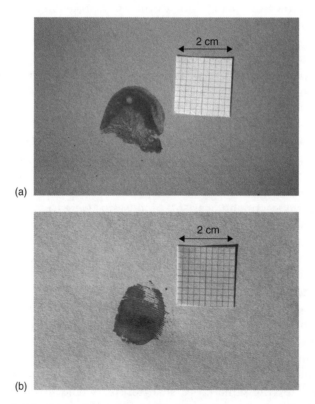

(a)

(b)

Figure 2.4 Transfer bloodstains from a bloody thumb formed on (a) a smooth plastic surface and (b) rough paper. Note the tendancy for smearing on the smooth surface. Square scale = 20 mm

by the passage of the head (especially if it has long hair) through them, e.g. if the victim is wounded in the chest and dragged by the feet. However, the direction of travel will remain obvious. A 'thinking murderer' wishing to move his victim's body would be expected to drag it by its feet because this reduces the risk of blood being transferred onto his clothes. If a dead body is held under the armpits and then dragged, a smear might be expected from one or both heels passing through the initial bloodstains.

If more than one person is present when a violent crime is committed, it may be possible to distinguish their roles from the types of bloodstains upon them. For example, the person pulling the trigger or wielding the murder weapon may have only blood spatter stains on his clothes whereas the one shifting the heavily bleeding corpse may acquire only transfer bloodstains. The absence of bloodstains on a suspect is not necessarily an indication of innocence because it is possible that he may have had time to wash and change his clothes. Alternatively, he may have removed his clothes before committing the crime or he may have been sufficiently far away from any flying blood not to be hit or shielded from it in some way.

The Sam Sheppard case

One of the best-known cases involving bloodstain pattern analysis occurred in 1954, in Cleveland, Ohio, in which Dr Sam Sheppard, a successful osteopath, was first convicted and subsequently exonerated of his wife's murder largely on the basis of evidence from bloodstains. It was well known that Dr Sheppard did not get on with his wife and therefore the strange tale he related following his wife's murder instantly aroused suspicions. His story began with him being woken by his wife's screams: he was sleeping downstairs on a couch. He said that he rushed upstairs but was knocked unconscious by an unknown 'bushy-haired' intruder. He then stated that he regained his senses outside the house (how he got there he did not know) and returned to find his wife dead. He then called his neighbour (the local mayor) before collapsing. When the police and the coroner eventually arrived they found Mrs Sheppard in the bedroom where she had been battered to death by more than 30 blows, the room was ransacked and a trail of blood could be followed down the stairs and out of the house. The coroner concluded that Dr Sheppard was responsible on the basis of bloodstains found upon a pillow and a bloody impression deemed to have been a transfer pattern of the murder weapon. He thought the weapon was probably a surgical instrument although what sort of instrument was never specified. The jury believed the coroner and Dr Sheppard was sentenced to life imprisonment.

At a subsequent retrial, Dr Paul Kirk, a Professor of Criminalistics at the University of California, Berkley, reassessed the evidence. Firstly, he noted that there were no bloodstains on the ceiling – this indicated that the weapon was swung from side to side rather than being brought down vertically on top of the victim's head. This was confirmed by the pattern of stains on the walls that could only have been caused by a combination of blood being flung outwards from the swinging murder weapon and direct contact between Mrs Sheppard's head and the walls. Despite the frenzy of the attack, certain regions of two of the walls remained free of bloodstains. In addition, blood spots on the right-hand side of Mrs Sheppard's bed were formed into streaks indicating that the blows were being delivered from left to right. Dr Kirk therefore placed the murderer as standing between the two beds in the room where his body shielded parts of the walls from the splattered blood. Furthermore, he thought that the murderer must have been using his left hand to wield the murder weapon. This would call into question Dr Sheppard's conviction because he was right-handed and not ambidextrous. To carry out such a sustained and brutal assault would have required more coordination than a person could be expected to possess when not using his normal hand. Even more importantly, the only bloodstains on Dr Sheppard were smears and streaks (not spatters) on his trousers and these could have been acquired by standing by his wife's bed in order to take her pulse. However, the amount of blood flying about the room during the attack would have left the murderer covered in gore. As far as the 'transfer stain' on the pillow was concerned, Dr Kirk demonstrated that this resulted from the folding of the pillowcase whilst it was still wet with blood. He thought that the murder weapon was much more likely to be a heavy metal torch. The laxness of the original observation was further demonstrated by the admission that the bloodstains had not even been subjected to rudimentary serological analysis. Dr Kirk's analyses were fundamental to Dr Sheppard being acquitted of his wife's murder after spending 10 years in prison. Nobody else has ever been charged with the offence.

Over 50 years later, this case continues to rumble on in the American courts as Dr Sheppard's estate (Dr Sheppard died in 1970) is attempting to bring a multi-million dollar claim for unfair imprisonment against the Ohio State. Needless to say, the scenario presented above is a much-simplified version of a case that still contains many unanswered questions and the forensic data have been subject to numerous re-interpretations. For example, one expert witness has suggested that the blows could have been delivered by someone lying or kneeling over Mrs Sheppard before, during or after committing a sexual assault (there is some evidence that this occurred). Similarly, there is a dispute over Dr Kirk's assertion that the assailant must have been left-handed, and the absence of bloodstains on Dr Sheppard's upper body could have been due to him changing his clothes before summoning help. For more details, see Chisum (2000) and www.courttv.com/archive/trials/sheppard/witness.html.

The Sion Jenkins case

On 15 February 1997 13-year-old Billie-Jo Jenkins was bludgeoned to death with an 18-inch iron tent peg (also described in press reports as a tent spike and an iron bar) outside her family home in Hastings, East Sussex. At the first trial, her foster-father, Sion Jenkins, was convicted of her murder and sentenced to life in prison. However, Mr Jenkins has repeatedly stated his innocence and the case contains many anomalies that were debated at length in two appeals and two further murder trials. The prosecution claimed that the distribution of 158 microscopic bloodstains found on Mr Jenkins' clothes was acquired when, for reasons that have never been satisfactorily explained, he attacked his foster-daughter with the tent peg, before driving off on an unnecessary shopping trip with his other daughters to provide an alibi. By contrast, the defence maintained that whilst Mr Jenkins was away on his shopping trip Billie-Jo was attacked by an unknown intruder who then fled before Mr Jenkins returned. In this scenario, the stains represent blood droplets exhaled from Billie-Jo's airways onto Mr Jenkins' clothing when he went to her aid.

One of the first points of dispute was why, if Mr Jenkins really was the killer, his clothing was stained with only microscopic spots of blood when Billie-Jo had been struck at least eight times with a heavy metal object in a frenzied attack and the surrounding area was coated with blood. Attempts to replicate the attack using a pig's head covered in blood always resulted in the striking arm becoming covered in bloodstains yet Mr Jenkins, who is right-handed, had only three microscopic spots of blood on his right arm. The prosecution have countered that this is not a serious issue and furthermore they claimed to have identified a white substance that they took to be human flesh on three of the bloodspots on Mr Jenkins' trousers. This was suggested to be evidence that Mr Jenkins had not only been spattered with blood but also with flesh – indicating that he was the person delivering the blows. It is not stated in the published reports what tests, if any, were done to prove that the white spots were indeed flesh and that they came from Billie-Jo and that they could not have been passively transferred.

In the first trial, the forensic scientists called by the prosecution claimed that in the scenario presented by Mr Jenkins Billie-Jo would have been dead by the time he arrived back from his shopping trip and therefore incapable of spluttering blood over him. If a person is decapitated there is little doubt about when they will cease to breathe, but on the basis of head injuries, even horrendous ones such as those received by Billie-Jo, it will remain a matter of conjecture. Not surprisingly, therefore, medical experts have failed to agree on this issue. The defence have argued that Billie-Jo was still alive – albeit only just – when Mr Jenkins arrived on the scene. Histological examination of Billie-Jo's lungs has supported Mr Jenkins' version of events by demonstrating evidence of interstitial emphysema and also that Billie-Jo could have been alive for at least 20 minutes after the attack. Interstitial emphysema is a condition in which air escapes through a tear or other injury into the connective tissue of the lungs. It is a common occurrence when the airways are obstructed (as is argued by the defence in the case of Billie-Jo) or there is a penetrating wound to the chest. It was highly unfortunate that at the initial postmortem a histological examination of the lungs was not performed even though this is a recommendation in the guidelines provided by the Home Office and the Royal College of Pathologists. It is apparently a common feature of pathologists in England and Wales to work on their own in murder investigations and this means that there may be no 'second opinion'. In Scotland, as is often the case, the laws are slightly different and forensic pathologists work in twos on murder cases. Such mistakes are therefore less likely to occur although it is inevitably more costly in terms of time and money to conduct investigations this way. The defence goes on to argue that when Mr Jenkins saw Billie-Jo he ran to her and placed his hand on her body – this slight movement was then sufficient to dislodge a blockage in her larynx caused by blood seeping from her head wounds. As a consequence of the blockage, the air in Billie-Jo's lungs would have been under pressure. By removing the blockage, the pressure would force air out of the lungs along with a fine mist of blood droplets. Hence anyone who was near to her mouth would have been coated with tiny blood particles without even noticing it. Further support for this scenario was provided in the most recent retrial when forensic scientists called by the prosecution admitted for the first time under cross-examination that it was a reasonable possibility that Billie-Jo had expired the blood droplets over Mr Jenkins. It is always possible that (as was suggested in the Sam Sheppard case) the lack of extensive bloodstains on his clothes and body might have been because he cleaned himself up after killing Bille-Jo. However, according to the defence counsel, the timings and witness statements would suggest that in this scenario Mr Jenkins would have had only about 3 minutes in which to get angry with Billie-Jo, kill her and clean himself up. He would then need to calm down and take his other daughters out shopping without them thinking he was unduly upset.

To add to the uncertainties associated with much of this case, at the second trial the QC acting for Mr Jenkins claimed that important evidence had gone missing. For example, only approximately 40 of the blood spots remained for future analysis and the hospital storing Billie-Jo's lung tissue samples appeared to have lost them. There are further strange features associated with this case that do not 'add up', such as the testimony of the other members of Mr Jenkins' family. There is also the mysterious finding of a scrap of black plastic bin liner that had been forced deep inside Billie-Jo's left nostril and the role (if any) of 'Mr X' – a paranoid schizophrenic who was known to have been in the neighbourhood and whom police had observed pushing plastic bags into his own mouth and nose. At the end of the second trial, the jury failed to reach a verdict and was discharged. Despite this, Sion Jenkins was immediately tried for a third time but this jury was also unable to reach a verdict and he was formally cleared of murdering Billie-Jo on 9th February 2006. More background information on the Billie-Jo Jenkins case, albeit biased towards the case of the defence, can be found at www.innocent.org.uk/cases/sionjenkins/.

The cases of Sam Sheppard and Sion Jenkins have been provided in some detail to illustrate how 'messy' and confusing the evidence can be in a forensic investigation and therefore how experts not infrequently arrive at completely different conclusions when analysing the same material.

Saliva and semen as forensic indicators

Both saliva and semen are products of exocrine glands (i.e. they are secretions that are released to the outside of the body through ducts) and both can provide valuable forensic evidence. Saliva is a slightly alkaline secretion that is produced constantly by the salivary glands and released into the mouth. Here it lubricates the membrane surfaces and aids the chewing and swallowing of food. Although it contains the enzyme salivary amylase that begins the breakdown of starch, saliva does not have an important role in the digestion of carbohydrates. In addition, saliva also contains other enzymes, such as lysozyme, some salts and mucin. The presence of saliva stains can be demonstrated by detecting amylase activity. This is best done with a specific assay method, such as ELISA (enzyme-linked immunosorbent assay), that can distinguish salivary amylase from other amylases, such as pancreatic amylase and bacterial amylase (Quarino et al., 2005). Saliva is important forensically because traces may be left at bite marks and in spit, from which it is possible to isolate DNA. Furthermore, a number of drugs (in particular those used illegally in sport) can be detected in saliva; mouth swabs can be used as a source of DNA, thereby removing the need for a suspect to provide blood for a DNA test.

Semen is gelatinous fluid produced by the combined secretions of the seminal vesicles, the seminiferous tubules, the prostate gland and the bulbourethral gland. The typical ejaculate consists of between 1.5 and 5 ml of semen and contains 40–250 million spermatozoa. In addition to the sperm, the ejaculate also contains the sugar fructose that serves as an energy store, citric acid, calcium and a variety of proteins. Semen analysis is usually a feature of cases of rape and sexual assault. Semen may also be released following death and should therefore not be assumed to denote recent sexual activity (Shepherd, 2003).

Both saliva and semen fluoresce in UV light and spots observed on the body or clothing are circled with an indelible marker for future extraction and confirmatory analysis. Commercial semen detection kits are available over the Internet for spouses and partners who suspect that their 'significant-other' may be cheating on them. These usually rely on detecting the enzyme acid phosphatase on stained clothing and the assumption that a positive reaction can only be explained by sperm belonging to another man (jealous male) or consorting with another woman (jealous female). It should be borne in mind that acid phosphatase is a very common enzyme and the test is indicative rather than proof of the presence of semen. Another, albeit more time–consuming, method is to look for the spermatozoa. These may be found within the vagina up to 5 days after sexual intercourse although the frequency of detection declines rapidly with time (Willott and Allard, 1982). Human spermatozoa are extremely small and soon lose their tails, so a sensitive cytological test is required. The most effective method is the 'Christmas tree test' (Allery et al., 2001), which uses nuclear fast red and picroindigocarmine dyes to stain the sperm red and green. More discriminative tests for semen have been suggested, such as detecting the prostate specific antigen (PSA) that is normally used in the diagnosis of prostate cancer (Maher *et al.*, 2002).

Identifying an individual from saliva or semen samples depends on either serological or DNA profiling. Approximately 75–85 per cent of the population are termed 'secretors', i.e. their body fluids include the same or similar profile of antigens, antibodies and enzymes that are found in their serum (although not the blood cells themselves, which are normally confined to the blood vessels). This can therefore, theoretically, enable an assailant to be identified by serological profiling from, say, a bite mark, spit or semen. However, the serological profile of the blood does not always match that of the body fluids and this reduces the effectiveness of this approach. DNA profiling offers more powerful discrimination but even this technique has its limitations. For example, in the UK, rapists are now aware of the risk of detection by DNA profiling and it is not unusual for them to use spermicidal condoms. It has been suggested that the use of such condoms (if found) might provide an indication of when the assault occurred. In an experimental study, the proportion of viable sperm within spermicidal condoms was found to decline gradually from about 40 per cent to 6 per cent over the course of 3 days (Gosline, 2005). This measurement would be useful if there was a dispute about whether the condom was 'planted' on the suspect or when the event took place. Although these findings are interesting, more work needs to be done on how sperm viability within the condoms is affected by temperature, light and other environmental conditions. Furthermore, many rapists fail to ejaculate (Scutt, 1990) or if they have had a vasectomy their semen lacks spermatozoa and hence DNA.

Despite the huge advances in forensic science, it is a sad reflection on justice in England and Wales that although the number of reported cases of rape has risen, less than 6 per cent of these result in a conviction (Home Office figures, 2002, http://newsvote.bbc.co.uk/mpapps/pagetools/print/news.bbc.co.uk/1/hi/uk/4296433.stm). This emphasizes that the successful resolution of a crime depends on more than the development of ever more sophisticated laboratory techniques.

Faeces and urine as forensic indicators

Faeces and urine are sometimes found at a crime scene because the culprit wishes to violate further the body or property of his victim. Similarly, during a violent attack, both the assailant and the victim can experience such stress and fear that defecation or urination is involuntary. Faeces and urine may also be recovered from a toilet that was not flushed and the body of a dead crime victim. Faecal analysis can indicate a person's diet and hence, possibly, where he or she had been eating. Some drugs and poisons can be detected in faeces and it can also be a source of DNA (Hopwood *et al.*, 1996). For example, in 1995, the murderer of Monica Jepson, a 66-year-old widow, left few clues behind at the nursing home in Birmingham where she was staying. Just about all the police had to go on were a partial fingerprint and some faeces left on the nursing home's fire escape. At the time, DNA profiling techniques were still in their infancy although it was possible to eliminate one suspect. By 2001, techniques had improved and a full profile was obtained. This was then compared to the National DNA Database© and a match was obtained with John Cook. His DNA was present on the database as a consequence of a previous offence and when brought in for questioning his fingerprints were found to match those left at the crime scene. He was subsequently charged with murder and sentenced to life in prison.

The study of DNA from faeces is difficult because it contains substances that interfere with standard DNA extraction and analytical techniques. For example, the human DNA needs to be differentiated from the large amounts of bacterial DNA, whilst the DNA amplification process can be inhibited by the presence of bile salts and plant polysaccharides (Lantz *et al.*, 1997). However, a commercially available kit for the extraction of DNA from faeces, the QIAGEN QIAamp® DNA Stool Mini Kit, is now available and is reportedly extremely effective (Vandenberg and van Oorschot, 2002; Johnson *et al.*, 2005).

Unless a person has a kidney or urinary tract infection, urine does not normally contain many cells although it is possible to extract DNA from urine and urine stains (Nakazono *et al.*, 2005). It is easier to obtain complete DNA profiles from women's urine than that of men because they shed more epithelial cells (Prinz *et al.*, 1993; Nakazono *et al.*, 2005) and also, presumably, because during the menstrual cycle their urine will contain blood cells. In forensic science, urine analysis is more frequently employed in the search for illegal drug use – especially drug abuse in sport. An interesting piece of research conducted by Zucatto *et al.* (2005) demonstrated the feasibility of detecting cocaine and its main metabolites in sewage water. They then scaled up their results to estimate the amount of the drugs flowing through the sewage water systems of four Italian cities. Their findings indicated a level of drug use far above that which the authorities believed to occur. For example, they estimated that the 5 million people living along the river Po were consuming about 200 000 lines of cocaine every day. It is possible that a similar approach might be used covertly to monitor levels of drug use in a house, nightclub or prison by analysing the drug levels in the sewage pipes.

For a summary of the forensic information that can be obtained from body fluids and waste products, see Table 2.2.

Molecular biology

The structure of DNA

DNA is quite literally 'the stuff of life' because it contains all the information that makes us who we are. Indeed, some biologists suggest that the sole reason

Table 2.2 Summary of forensic information that can be obtained from body fluids and waste products

Sample	Forensic information	Test
Blood, serological markers	Identification	Blood typing Rhesus factor Enzyme polymorphisms
Blood, molecular markers	Identification	DNA profiling
Blood, chemistry	Poisoning	Specific tests (e.g. carboxyhaemoglobin)
	Drug use	Specific tests (e.g. heroin metabolites)
Blood, infections	Previous travels	Specific tests (e.g. malaria, leishmaniasis)
	Source of infection (if being spread recklessly or deliberately)	Specific tests (e.g. HIV, hepatitis)
Bloodstains	Sequence of events during a violent assault	Bloodstain pattern analysis
	Linking an assailant to a victim and/or location	Bloodstain pattern analysis + DNA profiling
Saliva and semen	Identification	DNA profiling
	Linking an assailant to a victim and/or location	DNA profiling
Faeces and urine	Identification	DNA profiling
	Linking an assailant to a victim and/or location	DNA profiling
	Previous diet (and hence linking the individual to a location)	Undigested food residues in faeces
	Poisoning	Specific tests for the poison or its metabolites
	Drug use	Specific tests for the drug or its metabolites (e.g. morphine, erythropoietin)

for any organism's existence is to ensure that its DNA is replicated and survives into the next generation. It is found in the nucleus of cells and also the mitochondria. Therefore, with the exception of certain specialized cells that lack these organelles, such as mature red blood cells, every cell in the body contains DNA. Furthermore, this DNA is identical in every cell (unless heteroplasmy occurs; see later section on Mitochondrial DNA), does not change during a person's lifetime and is unique (identical twins excepted) to that individual. DNA is composed of four nucleotides (adenine, thymine, cytosine and guanine) and phosphate and sugar molecules. Nuclear DNA takes the form of a ladder twisted into the shape of a double helix in which the rails are composed of alternating sugar and phosphate molecules whilst the nucleotides act as rungs joining the two rails together. Adenine is always joined to thymine and cytosine is always joined to guanine.

Within the nucleus, DNA is found in structures called chromosomes. Human cells contain 23 pairs of chromosomes and these vary in shape and size. Twenty-two of these pairs are referred to as the 'autosomal chromosomes' and these contain the information that directs the development of the body (body shape, hair colour, etc.). The remaining pair of chromosomes are the X and Y 'sex chromosomes' that control the development of the internal and external reproductive organs. Each chromosome contains a strand of tightly coiled DNA. The DNA strand is divided into small units called genes and each gene occupies a particular site on the strand called its 'locus' (plural 'loci'). The total genetic information within a cell is referred to as its 'genome'. There are about 35 000–45 000 genes and, on average, they each comprise about 3000 nucleotides although there is a great deal of variation. These genes code for proteins that determine our hair and eye colour, the enzymes that digest our food and every hereditable characteristic. Surprisingly, only a small proportion of the genome actually codes for anything and between these coding regions lies long stretches of repetitive non-coding regions (so-called 'junk DNA') that exhibit a great deal of variability. Each gene exists in two alternative forms called 'alleles', one of which is found in each of the pair of chromosomes. If DNA profiling detects only one allele, this is usually interpreted as a consequence of a person inheriting the same allele from both parents. If three or more alleles are detected then this is an indication that the sample contains DNA from more than one individual. Mitochondrial DNA is arranged slightly differently to nuclear DNA and will be dealt with separately.

Because the sequence of nucleotides along the nuclear DNA chain is unique to every individual and is the same in every cell in his or her body, it is often stated to be similar to a 'bar code' for identification. Although the entire human genome has been sequenced, it is not necessary to go to these lengths in order to identify an individual. Indeed, the majority of the DNA in every one of us is virtually the same, so sequencing all of it would be a pointless exercise. Instead, forensic scientists concentrate on regions of the genome that exhibit a high degree of variability.

DNA profiling

Because virtually all of our cells contain DNA and we lose cells all the time, for example whenever we blow our nose, brush our teeth or comb our hair, it is

Table 2.3 Potential sources of human DNA for forensic analysis

Body fluids: blood, semen, saliva, urine, faeces, vomit
Tissues: skin, bone, hair, organs, fingernail scrapings
Fingerprints
Weapons
Bites
Discarded chewing gum
Cigarette butts
Handkerchiefs and discarded tissues
Used envelopes and stamps
Cutlery
Used cups, mugs and bottled or canned drinks
Clothing
Hairbrushes
Toothbrushes
Shoes and other footwear
Plasters
Used syringes

possible to isolate DNA from a wide variety sources (Table 2.3). Indeed, it is so easy to leave a trail of DNA that crime scene investigators must wear masks and disposable over-suits and over-shoes to avoid contaminating the location. Similarly, all DNA samples need to be kept apart from the moment they are collected, to avoid the possibility of cross-contamination. For example, if samples from a victim and a suspect are transported in the same container (even if they are in separate bags) or processed at the same time, there is always a risk of cross-contamination. The sequencing of an individual's DNA is known as 'DNA profiling' or 'DNA typing' and although the term 'DNA fingerprinting' is sometimes used it is not really appropriate. This is because the courts and the scientific community accept fingerprints as unique identifying features. For example, identical twins do not have the same fingerprints even though they share the same DNA profile. Consequently, experts presenting DNA-based evidence in court talk about probabilities of a match between two samples rather than stating 'yes, they match' or 'no, they do not match'. The possibility that 'an evil twin brother' committed the crime does not belong solely in the realms of fiction. Several cases have arisen in the UK and America in which DNA recovered from a crime scene could have been derived from either of two brothers. Often, these can be resolved by other evidence, such as fingerprints or one of the twins being in prison at the time of the offence. However, in the absence of such evidence and if the twins fail to cooperate with the police, it is an extremely difficult situation for the prosecution to resolve. It also indicates that DNA databases need to be updated so as to keep track of where twin matches could occur.

The Colin Pitchfork case

The first and most famous case of DNA profiling in a forensic investigation involves that of Colin Pitchfork, who in 1988 was sentenced to life in prison for the rape and murder of two young girls in the town of Narborough, Leicestershire (www.forensic.gov.uk/forensic_t/inside/news/list_casefiles.php?case=1). The first girl was raped and murdered in 1983 and a semen sample recovered from her body indicated that the culprit had blood group A and an enzyme profile shared by only 10 per cent of the male population. However, with no further leads, the investigation never progressed any further. In 1986, another girl was raped and murdered in the same area and the semen sample showed the same characteristics. These findings led the police to believe that both murders were the work of the same man and that he lived in the vicinity. A local resident, Richard Buckland, was subsequently arrested and whilst he confessed to the second girl's murder he denied having anything to do with the first girl's death. The police were convinced that they had arrested the correct person and asked Dr Alec Jeffries of Leicester University for help because he had developed a means of creating DNA profiles. Dr Jeffries' findings indicated that whilst both girls were raped and murdered by the same man, this could not have been Richard Buckland. This emphasizes the need for caution even when a suspect confesses to a crime.

With their chief suspect exonerated (another first for DNA profiling), the police undertook the world's first DNA mass intelligence screening in which all the men in the area, 5000 in all, were asked to provide DNA either as a blood sample or a saliva swab. Of these samples, only those exhibiting the same blood group and enzyme pattern as the murderer were subjected to DNA profiling. This was a major operation, not least because the profiling techniques were much more time consuming than those in use today and took 6 months to complete. At the end of this time, the operation drew a blank – none of the profiles matched those of the murderer. There followed a lull of about a year until a woman reported that she had overheard a man saying that he had provided a DNA sample in place of his friend Colin Pitchfork. Colin Pitchfork, who was a local baker, was therefore arrested and his DNA profile was found to match those of the semen samples recovered from the two murdered girls.

Forensic applications of DNA profiling

DNA profiling continues to be used for the identification of both victims and assailants in serious crimes, such as rape, sexual assault and homicide. However, as our understanding of DNA has increased and the technology improved, it is being used in a wider variety of situations. For example, it is now routinely used in minor crimes such as car theft and burglary and to identify the senders of hate mail and poison pen letters. It is used for paternity testing and many commercial firms now offer independent checks for anyone who wishes to check whether their offspring or parents really belong to them. There are serious ethical implications about this but these cannot be entered into here. DNA profiling is also used in insurance company fraud investigations where there is a dispute over who the driver was in a traffic accident. For example, when a car is in an accident and the airbags are activated, they deliver a considerable blow and this, coupled with the forces from the accident, means that the bags become contaminated with saliva, nasal mucus, vomit and, not infrequently, blood. Consequently, it is possible to obtain DNA profiles that can be used to demonstrate who was sat where at the time of the accident. Similarly, there are often claims that food has been contaminated with blood or other body fluids and it is not unusual for these to prove to have originated from the consumer. The disaffected employee responsible for the intentional contamination of food with spit, urine and other body fluids during processing in the factory or kitchen can also be identified by DNA profiling.

DNA profiling is not restricted to solving crimes involving human victims and Lorenzini (2005) relates an interesting case in which it was used to solve a case of wildlife poaching. A poacher had snared a wild boar in an Italian National Park and then killed it with a knife, leaving the corpse under a bush so that he could collect it after nightfall. Whilst he was away, conservation officers found the body and attempted to arrest the poacher when he returned. However, the poacher claimed that the boar was already dead when he found it and they had to let him go – though they confiscated the boar's body. A postmortem on the boar indicated that the knife wound was inflicted when the animal was still alive and a search of the poacher's home yielded a home-made knife that was covered in bloodstains. DNA analysis indicated that the blood originated from a wild boar rather than a domestic pig and that the DNA profile matched that of the dead boar. As a result, the suspect was successfully prosecuted for animal cruelty and poaching.

The huge popularity of TV dramas and documentaries featuring forensic science has led to concerns that the public will develop unreal expectations of forensic investigations. In particular, in the real world, investigations are seldom quick and they are also fallible. For example, during 2004–2005 some forensic laboratories in the UK and America developed backlogs of material owing to the increased volume of DNA tests that are now requested. Furthermore, if the DNA is badly degraded it may be impossible to obtain a good profile, or any profile at all. In addition, there are concerns that criminals will be alerted and learn how to avoid detection. For example, it has been reported that before abandoning a stolen car the thieves sometimes litter it with discarded cigarette butts so as to make it appear as though numerous people had used the vehicle.

Evaluation of DNA evidence

If the DNA profile of a suspect does not match that of the evidence, it is said to be 'an exclusion'. This is a useful result because it enables police to rule a suspect out of their enquiries or a convicted person to prove his or her inno-cence. If the interpretation of the DNA profile fails for some reason, such as owing to contamination or DNA degradation, the results are said to be 'incon-clusive'. This means that further evidence must be sought before a suspect can be either excluded or confirmed as being relevant to the enquiries. If the pro-files are deemed to match, then the suspect's profile is called 'an inclusion' and the significance of the match has to be calculated by quantifying the 'random match probability (Pm)'. This is a statistical test that estimates the chance of two unrelated people sharing the same DNA profile. Where the markers are not linked (e.g. autosomal short tandem repeats – see later), and are therefore inher-ited independently, the Pm can be estimated by multiplying the individual allele frequencies in a sample population (Jobling and Gill, 2004). Consequently, the more loci that are included in the analysis and the greater the heterozygosity of each of these loci, the smaller will be the value of the Pm and the greater will be the probability that the two profiles originated from the same person.

It is important to remember that despite the high discriminating power of DNA-based evidence, in the absence of other corroborating evidence it is unlikely to be accepted by a court as proof that a person was responsible for a particular crime (www.forensic-evidence.com/site/EVID/DNA_Watters.html). The Pm value can be compromised by several factors that need to be taken into account. Firstly, if the DNA has begun to degrade, it may not be possible to obtain a full profile that accounts for all the loci. Secondly, both the victim of a crime and the suspect may be closely related (many serious crimes are com-mitted by people related to their victim) and therefore share many alleles by descent. Similarly, they may both originate from the same sub-population, some of which are characterized by high levels of intermarriage.

Polymerase chain reaction

Most modern DNA profiling is based on the polymerase chain reaction (PCR). Kary Mullins invented the method in 1983 and it has since become one of the most powerful techniques in molecular biology. It is an enzymatic process that enables a particular sequence of the DNA molecule to be isolated and ampli-fied (copied) without affecting the surrounding regions. This makes it very useful in forensic casework in which DNA samples are frequently limited in both quan-tity and quality. For instance, PCR has been applied to the identification of DNA from saliva residues on envelopes, stamps, drink cans and cigarette butts (Withrow et al., 2003). It also has the advantages of being sensitive and rapid. However, PCR is not suitable for the analysis of long strands of DNA, so it cannot be used in the older restriction fragment length polymorphism (RFLP) analyses in which the strands often contain thousands of bases.

Once a region of the DNA molecule has been identified as being worthy of investigation, the flanking sequences are ascertained so that PCR primers can be designed to identify the beginning and end of the sequence. The primers consist of short sequences of DNA that bind or hybridize onto their complementary sequences on the test DNA sample. Once the primers have been designed, the PCR process is carried out as follows:

1.   Incubating the sample at 95–97°C to separate the DNA helix into two separate strands denatures the DNA.

2.   The temperature is reduced to 50–60°C to allow the primers to 'anneal' to the DNA.

3.   The temperature is raised to 70–72°C to inititate the 'polymerization' stage in which *Taq* DNA polymerase enzyme uses the DNA template identified by the primers and the nucleotides adenine, guanine, cytosine and thymine as building blocks to reproduce a complementary copy of the template.

4.   This procedure is repeated in successive cycles of denaturing, annealing and polymerization so that in a very short time the original sequence is 'amplified' thousands or even millions of times.

5.   After the amplification step, the PCR products are separated by electrophoresis on the basis of length. In the past this was done using flat-bed gel electrophoresis but this is being superseded by capillary electrophoresis because it is faster and can be automated. It is common practice to investigate several different sequences at the same time in a process referred to as 'multiplexing'. This is achieved by designing primers that produce allele size ranges that do not overlap.

Taq DNA polymerase is so called because it was discovered in the thermophilic bacterium *Thermus aquaticus*. This enzyme has replaced the *Escherichia coli*-derived DNA polymerase that was used in early PCR reactions because it is more heat stable. It is not harmed by the denaturation part of the PCR cycle, thereby removing the need to add fresh enzyme after each denaturation. Consequently, an excess of *Taq* DNA polymerase, primers and nucleotides is added at the start of the process and adjusting the annealing temperature controls the specificity of the reaction.

The primers are labelled with differently coloured fluorescent dyes and therefore, after electrophoresis, the PCR products can be detected by exposure to a laser beam that induces fluorescence at specific emission wavelengths that are then detected with a recording CCD camera. The results are printed out as a trace referred to as 'an electropherogram'. Sometimes the machine may misinterpret a colour (e.g. it mistakes blue for yellow) and this gives rise to false peaks – this phenomenon is called 'bleed through' or 'pull up'. This can be recognized by careful analysis of the electropherograms across the colour spectrum. Other potential sources of error are 'stutter peaks' that occur immediately in front of (commonly) or after (less commonly) a real peak. Stutter peaks are easy to identify and exclude from the interpretation when the sample is derived from a single person but if it contains mixed DNA it can be difficult to discern a stutter peak from a real one. In addition, random flashes may occur owing to air bubbles, contaminants and other interferences, thereby resulting in background 'noise' that may mask small peaks or even be mistaken for peaks themselves. Re-running the sample will usually identify 'false peaks' because they are unlikely to occur in exactly the same place twice. Surprisingly, there appear to be no universally accepted guidelines concerning the lower accepted limits that distinguish a 'true peak' from 'background'. When amplifying very low levels of the DNA template, further problems can arise through 'stochastic fluctuation'. This phenomenon occurs from the unequal sampling of the two alleles found in a heterozygous individual. For example, PCR reactions involving DNA template levels of about 100 pg DNA, or approximately 17 diploid copies of genomic DNA, can produce allele dropout. Consequently, a heterozygous individual may appear to be homozygous. This emphasizes the need for caution when interpreting the results rather than accepting them at face value – but this is true for any machine, whether it is a hand-held calculator or an automated PCR analyser.

Short tandem repeat markers

Short tandem repeats (STRs), also referred to as 'microsatellites' or 'simple sequence repeats (SSRs)', are brief lengths of the non-coding region of the human genome consisting of less than 400 base pairs (hence 'short') in which there are 3–15 repeated units, each of 3–7 base pairs (hence 'tandem repeats'). These STR sequences, or 'markers', can be divided into three categories: 'simple', 'compound' and 'complex'. Simple STRs are those in which repeats are of identical length and sequence units. Compound STRs consist of two or more adjacent simple repeats whereas complex STRs have several repeat blocks of different unit length and variable intervening sequences. The STRs occur on all 22 pairs of autosomal chromosomes and the X and Y sex chromosomes. They can vary greatly between individuals, this diversity resulting from the effects of mutation, independent chromosomal variation and recombination. However, STRs found on the Y chromosome exhibit less diversity than those on other chromosomes because they do not undergo recombination. Consequently, STR diversity on Y chromosomes results solely from mutation.

There are over 2000 STR markers suitable for genetic mapping but only a few of them are used routinely for forensic DNA profiling. In the UK the Forensic Science Service (FSS) currently uses 10 autosomal STR markers, whereas in America the FBI uses 13. The DNA profiles are then stored on computerized systems – the National DNA Database® in the UK and the Combined DNA Index System (CODIS) in America. The use of a standard set of markers and computerized systems facilitates comparisons between the DNA profiles of suspects, convicted offenders, unsolved crimes and missing persons. If all 10 (or 13) STR loci in two DNA samples are found to have identical lengths then this is compelling evidence that they originated from the same person. Commercial testing kits are available for these loci and the kits also include a marker at the amelogenin locus to enable sex determination. Amelogenin is a substance involved in the organization and biomineralization of enamel in developing teeth. In humans, the gene is expressed on both sex chromosomes but that on the X chromosome is six base pairs shorter than that on the Y chromosome. Consequently, following PCR and electrophoresis, males being heterozygous (XY) express two peaks (or bands) whilst females being homozygous (XX) express a single peak. The test is not foolproof and problems can arise if there is a deletion of the amelogenin gene on the Y chromosome – an important consideration in some ethnic groups, such as Malay and Indian populations (Chang *et al.*, 2003).

A DNA profile report will often take the form of a table of alleles and a hypothetical example obtained using the ProfilerPlus™ system is illustrated in Table 2.4. The numbers relate to the position of the alleles at each gene locus. For example, in suspect 1 the gene D3S1358 is heterozygous and expresses alleles 15 and 16, whilst in suspect 2 the gene is homozygous and allele16 is expressed on both chromosomes. Suspects 1 and 2 have different profiles to that found in the semen stain and are therefore classed as 'exclusions', whilst the profile of suspect 3 is the same as that found in the stain and is therefore 'an inclusion'. The statistical frequency of that combination of alleles occurring in the population is then calculated by reference to a sample population. For example, for the locus VWA, if 8 per cent of Englishmen expressed allele 15 and 21.6 per cent expressed allele 16, the frequency of this pair of alleles would be $2 \times 0.08 \times 0.216$ = 0.0346, or 3.46 per cent of the male English population. If the frequencies at all the loci are added together, then the frequency estimate for the whole DNA profile will be extremely small – perhaps one in a hundred million or more. Obviously, a great deal depends on the sample population used to generate the allele frequencies and corrections may need to be made to allow for this. For example, certain allele combinations will be common among close relatives, sub-populations or ethnic groups but might be rare in the population at large. Consequently, the frequency of a DNA profile may be extremely low as a national average but might not be so rare among family members or an ethnic group.

Because STR analysis relies on the identification of sequences that are much shorter than those required for RFLP analysis, it is less vulnerable to problems associated with DNA degradation. The STR analysis can therefore be effective on older body fluid stains or corpses at a later stage of decomposition than would be the case with RFLP analysis. However, DNA degradation can still present difficulties. For example, peak heights may be reduced, thereby making it difficult to distinguish them from background 'noise'. Indeed, some peaks may disappear entirely whilst others remain visible, thereby resulting in an inaccurate profile. Evidence of DNA degradation is often exhibited by a progressive

Table 2.4 Table of alleles illustrating the hypothetical DNA profile of a semen stain and that of three suspects (Amel = amelogenin test for gender; Sus = suspect number)

	Allele loci									
	D3S3158	VWA	FGA	D8S1179	D21S11	D18S51	D5S818	D13S317	D7S820	Amel
Stain	16, 18	16, 16	19, 25	13, 14	29, 30	17, 17	11, 11	10, 11	9, 10	XY
Sus 1	15, 16	16, 16	19, 25	13, 14	29, 30	14, 17	11, 11	10, 11	9, 10	XY
Sus 2	16, 16	15, 16	21, 23	14, 14	27, 28	17, 17	10, 11	8, 9	8, 9	XY
Sus 3	16, 18	16, 16	19, 25	13, 14	29, 30	17, 17	11, 11	10, 11	9, 10	XY

decline in peak height with increasing sequence length. This is because longer sequences are more vulnerable to the effects of degradation. If the profile contains DNA from more than one individual, this problem can be exacerbated because the two (or more) DNA samples may not degrade at the same rate or in an identical manner. If one of a pair of alleles fails to be recorded, this is referred to as 'allele dropout'. Consequently, a heterozygous individual may appear homozygous at one or more gene loci. This can also occur when insufficient DNA is available to analyse. Any DNA profiles obtained from degraded or extremely small DNA samples should therefore be treated with caution.

The shortness of the STR markers means that only small amounts of DNA are required (although if the amounts are extremely small problems can arise – see above). The marker sequences can be easily amplified using PCR and their shortness also reduces the risk of differential amplification. Because PCR amplification occurs in a non-linear manner, reproducibility is affected by stray impurities and the shorter the sequence, the less risk there is of this occurring. Moving the forward and reverse primers in as close as possible to the STR sequence can further reduce the size of the STR amplicons. This procedure is known as miniSTR analysis. Butler *et al.* (2003) have produced a set of miniSTR primers that allows the maximum reduction in size for all 13 CODIS STR loci and also several of those used in commercial STR kits. This approach is useful where there are problems with conventional STR analysis through allele dropout and reduced sensitivity of larger STR alleles (Schumm *et al.*, 2004).

Y-Short tandem repeat markers

Over 200 STR markers have been identified on the human Y chromosome. Between nine and eleven of these markers are used routinely in forensic science and commercial kits are currently available for at least six of them (Sinha *et al.*, 2003). They are particularly useful in cases of rape and sexual assault where there are mixed male and female DNA profiles and therefore separating the two is a major challenge. Unlike conventional STR analysis, there is typically only one peak or band for each STR type in Y-STR analysis and these can only originate from DNA from a male. In the case of multiple sexual assaults more peaks will be found, depending upon the number of men involved. The simultaneous detection of multiple Y-STR loci produces additional genetic information without consuming additional DNA (Tun *et al.*, 1999). A further advantage of Y-STR analysis is that it enables DNA profiles to be made in cases of sexual assault in which the man did not produce sperm owing to a medical condition or being vasectomized. In these circumstances, the absence of sperm would mean that only a very small amount of male autosomal DNA would be present and the female's autosomal DNA would swamp this. By specifically targeting the Y-chromosome STR markers it would be possible to target the minute amount of male DNA that would be present.

Because Y-STR markers exhibit much lower variability than autosomal STR markers, their discriminatory power is much less and unless a mutation has occurred all male relatives – sons, fathers, brothers, etc. – will share the same profile. This needs to be taken into account when assessing the strength of the evidence and could be a major problem when the suspect comes from an inbred population or a criminal family. However, like other DNA profiling techniques, their value can be as great in excluding suspects as in identifying a culprit.

Single nucleotide polymorphism markers

Single nucleotide polymorphisms (SNPs) arise from differences in a single base unit (Figure 2.5) and are the commonest form of genetic variation. They are found throughout the genome, including the X and Y sex chromosomes. Everyone has his or her own distinctive pattern of SNPs and this, therefore, provides a means of identification. Using a technique called mini-sequencing, the base at a given SNP can be determined and once the bases at several sites at different loci are known one can produce a profile similar to that of an STR profile. Using allele frequencies for each SNP, the likelihood of two persons sharing the same SNP profile can be estimated. Because the maximum number of alleles at each site is only four (A, C, G or T), 50–100 SNPs need to be examined to achieve the same discriminatory power as STR-based profiling (Gill, 2001; Gill *et al.*, 2004). However, a process called microarray hybridization allows numerous SNP loci to be examined simultaneously, so it does not take long. The stability of SNPs, compared to STRs, means that they are less likely to be lost between generations and they are sometimes used in paternity cases. However, a statistical simulation study by Amorim and Periera (2005) indicated that relying exclusively on SNP analysis would result in more inconclusive results than STR analysis.

Because SNP analysis requires minute quantities of sample and the segment size can be even smaller than that needed for STR analysis, the technique can provide information even when the DNA is severely degraded. However, its effectiveness is compromised in mixed DNA samples because it could be difficult to distinguish which SNP belonged to which person. Furthermore, a quantitative test would be required in this context and this is not possible with some

TTGACGT
AACTGCA

TTAACGT
AATTGCA

Figure 2.5 Diagrammatic representation of a single nucleotide polymorphism

of the current SNP assays. There would be less of a problem if the mixture were composed of DNA from a single male and a single female because Y-linked SNPs could only originate from the male. However, if more than one male could have contributed to the DNA sample or the sexes were the same (e.g. male rape), their separation could be difficult. A possible solution to this problem would be to identify tri-allelic SNPs (Phillips *et al.*, 2004), although according to Brookes (1999) these are 'rare almost to the point of non-existence'.

A commercially available test for ancestry has been developed called DNAW-itness[TM] that uses 176 SNPs to determine the percentage of European, East Asian, Native American and sub-Saharan African BioGeographical Ancestry (BGA) in a DNA sample (www.dnaprint.com). The test uses reference DNA samples collected from these geographical regions and an unknown sample is screened against them using a statistical procedure called 'maximum likelihood estimation'. The apparently odd (to European eyes) inclusion of 'Native American' in the BGA list is because this test was developed with an eye to the American genealogy market and many Americans would dearly like to have 'proof' of Native American ancestry. The technique has its limitations and is complicated by the increasing mobility of our societies and the frequency of intermarriage. For example, if the result of the DNA analysis was 70 per cent European and 30 per cent sub-Saharan African BGA the profile would be indicative of a person of either North African descent or the progeny of a Caucasian and an African American / European. However, it can be helpful when attempting to put a 'face' to a culprit because even eyewitness statements can be unreliable. Indeed, it is often stated in forensic literature that about 50 per cent of eyewitness identifications are incorrect. A classic example of this was the case of Derek Todd Lee, who was found guilty of the rape and murder of seven women in Louisiana, USA, in 2003. Although the police had a record of the culprit's DNA from the crime scenes, this did not match any of those stored on the FBI's CODIS database. In addition, the DNA tests used for the CODIS database do not provide any information on racial origin. The police were therefore reliant on eyewitness descriptions – and these indicated that the culprit was a White male. It therefore came as a surprise when subsequent tests using DNAW-itness[TM] indicated that the culprit's profile was 85 per cent sub-Saharan and 15 per cent Native American BGA. This information allowed the police to broaden the investigation and follow new leads. After about 2 months, the police arrested Derek Todd Lee whose genetic heritage matched that indicated by the SNP profile and whose STR profile proved to match that from the crime scenes (www.dnaprint.com).

Mobile element insertion polymorphisms

An alternative to SNPs for the identification of racial characteristics from DNA is the analysis of mobile element insertion polymorphisms based on short interspersed elements (SINEs). The commonest class of these are the so-called *Alu* elements, which are about 300 nucleotides long. Most *Alu* elements are fixed at a particular locus but a few subfamilies are polymorphic for insertion presence/absence and can be used to determine genetic relationships between populations (Watkins *et al.*, 2003). Ray *et al.* (2005) undertook a blind study using 100 *Alu* insertion polymorphisms and was able to use them to correctly infer 18 individuals as being of African, Asian, Indian or European descent. The SINE analysis requires standard PCR laboratory equipment but suffers from the drawback of being time consuming if manual systems are used. On the positive side, unlike SNPs, *Alu* insertions are not at risk of forward and backward mutations and it is possible that fewer insertions will need to be analysed in future once the most reliable ones are identified. In addition, gender can be determined by analysing *Alu* insertions (Hedges *et al.*, 2003) and this is therefore a useful method to employ if it is thought that amelogenin gene deletion had occurred on the Y chromosome.

Mitochondrial DNA

Mitochondria are intracellular organelles that generate about 90 per cent of the energy that cells need to survive. The numbers of mitochondria found in a human cell depend upon its energy needs and vary from zero in the mature red blood cell to over 1000 in a muscle cell. They are thought to descend from bacteria that evolved a symbiotic relationship with pre- or early eukaryotic cells many hundreds of millions of years ago. With time, the symbiotic relationship became permanent but the legacy is reflected by present-day mitochondria retaining their own bacterial-type ribosomes and their own DNA (referred to as mtDNA) that is distinct from that found in the cell nucleus. Each mitochondrion contains between 2 and 10 copies of the mtDNA genome. The inheritance of mtDNA also differs from nuclear DNA in that it is exclusively generated from the maternal side. This is because the sperm head is the only bit of a spermatozoon that enters the egg at the time of fertilization. Usually, the spermatozoon's tail and the mid piece (which is the only bit containing mitochondria) shear off as the head enters the egg's perivitelline space. Occasionally, a few mid-piece mitochondria are incorporated at fusion but these are subsequently destroyed by the egg (Manfredi *et al.*, 1997). Consequently, the only mtDNA present in the developing embryo is that derived from the egg and therefore as usual (it is alleged) the workforce is exclusively female.

Human mtDNA takes the form of a circular DNA molecule that contains 16569 base pairs. These code for 37 genes that in turn code for the synthesis of 2 ribosomal RNAs, 22 transfer RNAs and 13 proteins. Unlike nuclear DNA, the mitochondrial genome is extremely compact and about 93 per cent of the DNA represents coding sequences. The remaining, non-coding region is called the control region or displacement loop (D-loop). The D-loop region consists of about 1100 base pairs and it exhibits a higher mutation rate than the coding region and about 5–10 times the rate of mutation within nuclear DNA. The mutations occur as substitutions in which one nucleotide is replaced by another one: the length of the loop region is not changed. The mutations result from mtDNA being exposed to high levels of mutagenic free oxygen radicals that are generated during the mitochondrion's energy-generating oxidative phosphory-lation process. The substitutions persist because mtDNA lacks the DNA repair mechanisms that are found in nuclear DNA. These mutations result in sequence differences between even closely related individuals and makes analysis of the D-loop region an effective means of identification (Budowle *et al.*, 1999). Because the mtDNA is inherited only from the mother, it also allows tracing of a direct genetic line. Furthermore, unlike the inheritance of nuclear DNA, there are no complications owing to recombination.

The D-loop is divided into two regions each consisting of about 610 base pairs, known as the hypervariable region 1 (HV1) and hypervariable region 2 (HV2). It is these two regions that are normally examined in mtDNA analysis by PCR amplification using specific primers designed to base pair to their ends. This is then followed by DNA sequence analysis. Because of the high rate of substitutions, it is possible to analyse just these short regions and still differen-tiate between closely related sequences. It has been estimated that mtDNA may vary by about 1–2.3 per cent between unrelated individuals (Inman and Rudin, 1997) and although mtDNA sequencing does not have the discriminating power of STR DNA profiling it can prove effective where STR DNA analysis fails.

The mtDNA sequence of all the mitochondria in any one individual is usually identical – this condition is referred to as 'homoplasmy'. However, in some people, differences in base sequences are found at one or more locations. These differences arise from them containing two or more genetically distinct types of mitochondria. This condition is known as 'heteroplasmy' and it can have a sig-nificant impact in forensic investigations (Lo *et al.*, 2005). Heteroplasmy used to be considered relatively rare but it is now believed to occur in 10–20 per cent of the population (Gibbons, 1998). To make matters worse, it is now apparent that heteroplasmy is not necessarily expressed to the same extent in all the tissues of the body. For example, two hairs from a single person might have different proportions of the base pairs contributing to the heteroplasmy and this might result in an exclusion rather than a match (Linch *et al.*, 2001). This is because heteroplasmy may result from the high mutation rate or from either inheritance at the germ line level or the level of somatic cell mitosis and mtDNA replication.

In forensic science, mtDNA analysis is most frequently used where the samples do not contain much nuclear DNA – for example, a fingerprint or a hair shaft – or where the DNA has become degraded through the decomposi-tion process or burning (Bender *et al.*, 2000). Because there are numerous mito-chondria in a single cell and each mitochondrion contains multiple copies of the mitochondrial genome, it is possible to extract far more mtDNA than nuclear DNA. Epithelial cells, which are the commonest cell type used in forensic casework, contain an average of 5000 molecules of mtDNA (Bogenhagen and Clayton, 1974). Mitochondrial DNA analysis does, however, suffer from a number of problems. For example, all maternally related individuals are likely to have the same mtDNA sequences, so the discriminating powers are limited compared to autosomal STR analysis. Heteroplasmy can be considered either a problem or a useful trait depending on the circumstances. It can create problems because the mixed sequence is also what would be expected if there were more than one individual contributing to the DNA profile. A difference of only one base pair between the mtDNA profile of the sample and the suspect is considered insufficient to prove either a match or exclusion whereas a difference in two or more base pairs is grounds for exclusion. By contrast, heteroplasmy can provide an identifying characteristic where the suspect expresses the same heteroplasmy characteristics as the sample.

Other common problems associated with mtDNA analysis are that detecting differences in sequences is more time consuming and costly than determining differences in lengths – as is accomplished using STR analysis. In addition, the rarity of mtDNA sequences has to be determined by empirical studies and the results are not as statistically reliable as those for other types of analysis. Finally, owing to the high copy number per cell there is always a risk of contamination and cross-contamination associated with mtDNA sequencing.

For a summary of the advantages and disadvantages of the must commonly used methods of forensic DNA profiling, see Table 2.5.

The Tsar Nicholas II case

Towards the end of the Russian Revolution, on the night of either 16 or 17 July 1918, the Tsar and his family were shot and their bodies disposed of. For many years, it was believed that after they were shot their bodies were butchered, covered in sulphuric acid, burnt and thrown down a nearby mineshaft. However, there were also frequent rumours about various members of the family escaping and their fate was shrouded in mystery. Following the collapse of Soviet Russia, a previously secret report became public knowledge and it transpired that although the bodies were thrown down a mineshaft they were subsequently retrieved and buried in a concealed pit about 12 miles north of the town of Yekaterinberg. In 1991, the pit was located and the skeletons of five females and four males were retrieved from it. These were believed to include those of the Tsar, his wife Tsarina Alexandra and three of their daughters. The morphology of the skeletons indicated that those of Prince Alexei and Princess Maria were missing. This agreed with the contents of the secret report, which stated that after the family was shot two of the bodies were burned. However, it still needed to be determined whether the remains included all the other members of the royal family.

Nuclear DNA and mtDNA testing confirmed that a family group was present in the grave. Further mtDNA analysis indicated a match between the profile of

Table 2.5 Advantages and disadvantages of the most commonly used methods of forensic DNA profiling

Genetic marker	Advantages	Disadvantages
Autosomal STRs	Small sample size Slight DNA degradation not a problem Excellent discrimination	Discrimination seriously reduced if DNA badly degraded
Autosomal SNPs	Extremely small sample size Discrimination possible with badly degraded DNA Tests for ancestry available	Discrimination power lower than for STRs Ability to distinguish mixed DNA profiles lower than STRs
Y-linked STRs	Male specific therefore useful for mixed gender DNA samples Small sample size Slight DNA degradation not a problem	Low discrimination especially among male relatives Discrimination seriously reduced if DNA badly degraded
Y-linked SNPs	Male specific therefore useful for mixed gender DNA samples Very small sample size Slight DNA degradation not a problem	Low discrimination Difficult to distinguish individuals if DNA from more than two males present
SINEs	Gender determination not affected by allele deletion Tests for racial origin possible	Limited number of studies to confirm effectiveness
Mitochondrial DNA	Very small sample size Effective with even badly degraded DNA Heteroplasmy may facilitate identification	Lower discrimination than STRs Limited discrimination if individuals are maternally related Heteroplasmy may prevent accurate identification

the remains presumed to be those of the Tsarina and those of Prince Philip, Duke of Edinburgh. Because Prince Philip's maternal grandmother was the Tsarina's sister, he had inherited the same mtDNA profile. Mitochondrial DNA analysis of the bones thought to belong to the Tsar proved more complex but also even more interesting. Profile matches were found between these remains and those of two distant maternal relatives except for position 16169. At this point, the bones expressed a cytosine (C) nucleotide but that of the relatives expressed a thymine (T). Cloning experiments on the bones demonstrated the presence of a mixture of mtDNAs that differed only at this single position, the C form accounting for 70 per cent of the mtDNA. This suggested that the Tsar exhibited heteroplasmy (Gill *et al.*, 1994).

Although the results indicated that the bones were 98.5 per cent certain to belong to the Tsar, this was not sufficient for some commentators who suggested that there were problems with sequence background effects or contamination. The Russian Orthodox Church had canonized Tsar Nicholas II, so a great deal depended upon identifying the remains with absolute certainty because they would become objects of veneration and pilgrimage. Eventually, the mystery was solved when permission was gained to open the coffin of the Tsar's brother, Grand Duke Georgij Romanov, who had died of tuberculosis in 1899. Mitochondrial DNA analysis of the Duke's bones demonstrated an identical mtDNA profile to those presumed to belong to the Tsar – including the same C and T heteroplasmy at position 16169. This was the first instance in which heteroplasmy was used to facilitate human identification. The results, together with morphological studies on the bones and all the other DNA studies, meant that the likelihood ratio for the remains' authenticity was in excess of 100 million (Ivanov *et al.*, 1996).

Quick quiz

1. Briefly explain the uses and limitations of red blood cell antigens for identifying an individual.

2. In relation to bloodstains, distinguish between the appearance and causes of cast-off stains, impact spatter and transfer stains.

3. Why is it difficult to extract human DNA from faeces?

4. List ten common sources of DNA that you might find at a crime scene.

5. In relation to DNA profiling results, distinguish between 'an inclusion', 'an exclusion' and 'an inconclusive'.

6. Explain the purpose of the amelogenin gene test and also its limitations.

7. Explain the value and limitations of Y-STR markers for identification.

8. Distinguish between STRs and SNPs and their advantages and limitations for identification purposes.

9. In relation to mitochondrial DNA, explain what is meant by the term 'heteroplasmy' and how it can be both a problem and an advantage for identification purposes.

Project work

Title

Environmental factors affecting the rate of degradation of DNA in bloodstains.

Rationale

It has been suggested that the degree of DNA degradation may reflect the length of time since death in human tissues. Owing to their size and complexity, tissues may not be the ideal candidates with which investigate this relationship. Bloodstains, by comparison, are simpler, easier to work with and, although there are a variety of methods available to determine how long a person has been dead, there are no reliable means of ageing bloodstains.

Method

Fresh blood would be collected from an abattoir and used to prepare bloodstains of varying sizes on different surfaces. The stained items would then be left under different environmental conditions (e.g. low and high temperature, dark and light, high and low humidity) and at varying time intervals the amount of DNA degradation that had occurred would be determined using different marker systems such as autosomal STRs, Y-linked STRs, mtDNA and SNPs.

3 Human tissues and wounds as forensic indicators

Chapter outline

The outer body surface: skin; fingerprints; lip prints and ear prints; retinal and iris scans; tattoos

Hair: characteristics; assimilation of poisons, drugs and explosives

Bones: characteristics; determination of gender; determination of ethnic origin; determination of stature; determination of age; facial reconstruction; determination of age of remains

Teeth: characteristics; determination of gender and ethnic origin; determination of age; identification based on dental characteristics; effects of drug abuse on dental characteristics

Wounds: definitions; bruises; abrasions; lacerations; incised wounds; stab wounds; bone damage; gunshot wounds; bite marks; burns and scalds; postmortem injuries

Objectives

Describe the different ways in which fingerprints are formed and the methods for their detection.

Explain how hair can provide information on a person's identity and their exposure to chemicals such as methodone and explosives.

Compare the strengths and limitations of bones and teeth in determining a person's personal characteristics.

Critically evaluate the techniques available to determine the age and geographical origin of skeletal remains.

Discuss how wound analysis can be used to differentiate between suicide, accidental death and homicide.

Compare and contrast the wounds produced by low- and high-velocity firearms and how one can determine the distance from which a victim was shot.

Describe how postmortem injuries may be distinguished from those that occurred before or at the time of death.

Essential Forensic Biology, Alan Gunn
© 2006 John Wiley & Sons, Ltd

The outer body surface

Skin

Skin covers the whole surface of our body and is the largest organ in terms of both weight and surface area. It varies in thickness from about 0.5 mm on the eyelids to 4 mm or more on the heels. Repetitive abrasion of the skin surface results in the formation of calluses, the distribution of which, along with the condition of the fingernails, can provide an indication of a persons trade or activities. For example, manual labourers develop large thick calluses over their palms and fingers and their fingernails become chipped and ragged. At the opposite extreme, musicians develop calluses on highly localized regions and string instrument players may allow certain fingernails to grow long to facilitate plucking. Violinists typically develop calluses on the dorsal surface of their left second and third fingers over the proximal interphalangeal joints – these calluses have their own name, 'Garrod's pads'. Similarly, wind players develop calluses on the mid portion of the upper lip.

Structurally, skin can be divided into two regions, the outer epidermis and the thicker inner dermis. The epidermis is composed of several layers of cells that are often described as being a keratinized stratified squamous epithelium. Roughly translated, this means that the cells contain the fibrous protein keratin, present the appearance of a wall of bricks (stratified), are flat (squamous) and they form the outer surrounding layer (epithelium). Embedded within the epidermis are melanocytes: round cells with long, slender projections that contain the pigment melanin and are responsible for giving skin its coloration. Beneath the epidermal layers lies the dermis, which is largely composed of connective tissue but also contains blood vessels, nerve endings, glands and hair follicles.

Fingerprints

The deepest layer of the epidermis is known as the basal layer. In the 1920s, Kristine Bonnevie (the first female professor in Norway) suggested that when a developing human embryo is about 10 weeks old, the basal layer at the tips of the fingers grows faster than the surrounding upper epidermal layers and the lower dermal layers. This would result in the basal layer being stressed and thrown inwards into a series of folds that are manifested on the surface as 'friction ridge skin'. Similar skin is also found on the palms of the hands, the soles of the feet and the lower surface of the toes. The patterns of loops, whorls and arches that are formed are unique to every person. This hypothesis for how friction skin is formed has never been conclusively proven but computer-modelling studies by Kücken and Newell (2004) have lent it strong support. Regardless of the precise mechanisms by which they are formed, once we are born we already have our own unique fingerprint pattern that will remain the same for the rest of our life. As we grow bigger, all that happens is that the fingerprint ridges become further apart – indeed, there is a relationship between height and ridge width. After minor cuts, burns or bruises have healed, the normal fingerprint pattern is restored. Deeper injuries, such as serious burns, result in the formation of permanent scar tissue – and this is also a good identifying characteristic. Cosmetic surgery cannot erase the fingerprint pattern because to be effective it would require the removal of so much tissue that a person would be unable to use his hands effectively.

The study of fingerprints is known as dactyloscopy. It has been used as a forensic tool since the nineteenth century and as early as 1906 a conviction was upheld in England solely on the basis of fingerprint evidence (www.forensic-evidence.com/site/ID/Buckley_UK.html). Even identical twins do not have the same fingerprints and courts therefore accept them as unique identifying characteristics. Fingerprints are normally divided into three classes: plastic, visible and latent. Plastic fingerprints are those that are formed when a person touches a soft or semi-solid substance such as soap or unset putty (Figure 3.1). This results in a shallow three-dimensional record of the friction ridges that is either transient or long-lasting depending on the substance and the circumstances. For example, plastic fingerprints can be identified from 2500-year-old clay seals recovered from the ruins of the Mesopotamian city of Ur. Visible fingerprints, as their name suggests, are those that can be seen without further enhancement (Figure 3.2). They result from a combination of sweat and the oily secretions from the skin glands being deposited on a contrasting surface, thereby making the print visible. Alternatively, they may be formed when a hand covered with substances such as blood, ink or engine oil is pressed against a contrasting surface. Latent fingerprints are those that need to be enhanced in some way before they become visible (Figure 3.3). This is because sweat and body oils are colourless, so unless the prints are on a contrasting surface they cannot be seen

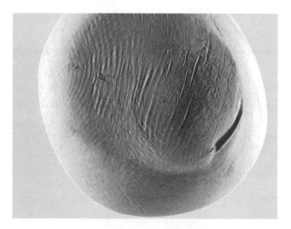

Figure 3.1 Plastic fingerprint formed in plasticene. Note the fingernail impression

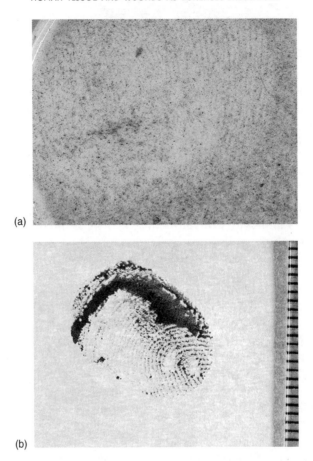

(a)

(b)

Figure 3.2 Visible fingerprint left on the dusty surface of a jar (a) and as an ink stain on paper (b). Note how fingerprints are not always clear or complete. Transfer fingerprints, such as (b), tend to smear if the fingers are 'overloaded' with liquid, even when placed on an absorbent surface

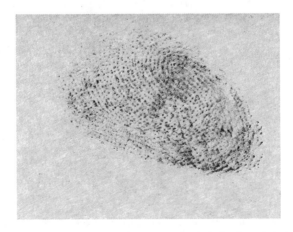

Figure 3.3 Latent fingerprint left on paper revealed using ninhydrin spray

easily. Washing one's hands before handling an object reduces the quality of latent fingerprints because it removes the sweat and body oils. Similarly, young children tend to leave poor quality fingerprints that degrade quickly because they do not produce as much body oil as adults. Latent prints may also result from the removal of a layer of material, such as fine dust, thereby leaving a 'negative impression'.

Searching for fingerprints is a specialist skill and should be undertaken with extreme care owing to the ease with which prints may be damaged or lost if the incorrect method is chosen. To begin with, all the visible prints must be located by passing a strong beam of light over the scene or object. The angle of the beam of light should be altered to make the prints 'stand out'. Sometimes latent fingerprints can be rendered visible simply by viewing them under different lighting conditions – oblique lighting is often effective. However, the use of special fingerprint powders or sprays is usually required, the choice of which depends on the nature of the print, the substance it occurs on and the personal preferences of the investigator. Powders stick to the moist or oily parts of the print and are employed on non-absorbent surfaces such as plastic or glass and they tend to be brushed on. 'Magnetic' powders are useful on textured surfaces, such as leather. These powders are not themselves magnetic but they contain iron and are applied with a magnetic wand. Magnetic attraction therefore sticks the powder to the wand and this is dragged slowly across the surface of an object. The components of fingerprints exert a stronger attractive force than the magnet, so the particles remain behind and thereby render the prints visible.

There are various means of chemically identifying fingerprints and the choice depends on the surface and whether the objects are required for other forensic examinations – some tests can interfere with one another. One of the most commonly used reagents is ninhydrin – this reacts with the amino acids found in sweat to form a pink to reddish purple-coloured compound. The ninhydrin reagent may be sprayed or brushed on, or even used as a dip. This method is especially effective for prints found on paper and other porous surfaces. It is possible to speed up the process of development of the prints by conducting the procedure at 50–70°C and 60–80 per cent relative humidity. This is known as accelerated ninhydrin process fingerprint development.

Silver nitrate is useful for identifying latent prints on brass objects such as ornaments and ammunition cartridges, but also porous surfaces such as wood. The silver nitrate reacts with sodium chloride in sweat to form silver chloride: this is light sensitive and under the influence of UV light decomposes to silver and chlorine. Consequently, the print becomes visible as a greyish or brownish stain, although this fades with time. Unfortunately, silver nitrate will react with any source of sodium chloride and this may obscure the print. Consequently, it would not be a good method for detecting prints on the brass fittings found on a sea-going yacht. Iodine fumes react with the oily components of latent fingerprints, thereby making them visible as a brownish colour. However, iodine prints are unstable and must be photographed immediately or 'fixed' in some way to preserve them.

Exposing latent fingerpints to ethyl or methyl cyanoacrylate (superglue) vapour renders the prints visible through the formation of a white deposit. The prints can then be enhanced by the application of fluorescent dyes or by observing the prints in UV light. This method has the advantage of being simple and cheap and works best for prints on glass, china and other non-porous surfaces. It should be done at room temperature and in a closed container, partly to protect the operator from the fumes but also to ensure that a high humidity of about 80 per cent is maintained. In addition to these methods, there are many other chemicals that can be used to render fingerprints visible. Sometimes a print may be subjected to more than one detection method in order to obtain as accurate a record as possible.

Once a print has been located and rendered visible, it should be photographed. The print may then be left *in situ* or lifted and stored separately from the object it was found on. If it is to be preserved on the object, the print should be protected by clear tape, varnish or lacquer. There are several means of lifting a fingerprint but it is best to use special fingerprint tape as this has a very smooth adhesive surface. The tape is pressed against the print and then lifted, after which it is pressed into a 'lift card'. Lift cards have a smooth surface and are usually transparent. The prints may then be photographed again and stored.

Fingerprints can be classified according to their distinctive patterns of loops, arches and whorls. However, identifying a fingerprint is not easy and relies on noting characteristic features and their spatial relationship to one another. There is little consensus among experts as to how much similarity between two sets of prints is necessary for them to be considered a perfect 'match'. In the UK a 16-point minimum was required for many years, whilst in France it was 17 and in the USA each state set its own standards. With time, it was accepted that there is little scientific or statistical basis for any numerical standard and there are strong arguments for moving to a non-numerical system (www.forensic-evidence.com/site/ID/Buckley_UK.html). There are now computer programs that speed up the matching process. For example, in England and Wales, the police forces use the National Automated Fingerprint Identification System (NAFIS) – this includes a database that currently contains over 8 million sets of prints and over 1 million marks from crime scenes. In America, the FBI operates a similar system called the Integrated Automated Fingerprint Identification System (IAFIS).

Changes to the law introduced in the Criminal Justice Act 2003 now allow police officers in England and Wales to take both DNA and fingerprint evidence from a person after they are arrested and detained at a police station. Previously, this could only be done after a person was charged with a criminal offence. The fingerprints can now be taken electronically using Livescan fingerprint technology and digital images can then be entered instantly into NAFIS. Consequently, within a few minutes, it is possible to check whether a person is lying about their identity or their prints match those recovered from a crime scene. Similarly, prints recovered from a crime scene can be compared with those on the database. However, the computer program's main purpose is to speed up a search: final confirmation of a match always relies on physical observation of the prints by a fingerprint officer.

Although it is generally accepted that everyone has their own unique finger-prints (the chances of two persons having identical fingerprints is usually stated as 1 in 64 billion), correlating the prints found at a crime scene with those of a suspect is not a simple matter. For example, the prints may be smeared, poorly formed or 'partial prints' such as half a thumbprint. Furthermore, prints decay with time at a highly variable rate that depends upon what they are composed of, the matrix they are formed on and the environmental conditions. For example, the lipid components of latent fingerprints can change dramatically with time (Archer *et al.*, 2005) and this should be borne in mind when decid-ing which enhancing agent should be used to render the prints visible. There is currently no accepted method for ageing fingerprints. Consequently, finding a suspect's fingerprints on a gun would indicate that he had handled it but not when he had handled it. This can be important when attempting to prove that a suspect committed a murder or was present in a room but he claims to have picked up the weapon or visited the room at an earlier date to the crime being committed.

Lip prints and ear prints

The use of lip prints for identification purposes is occasionally alluded to in forensic literature but there have been no published studies that have verified their individuality and there appear to be no accepted standards or methods for their identification and no databases. It would therefore be an interesting topic for students to investigate. Lip prints are currently most useful as a potential source of DNA from which a much more accurate identification may be obtained. There is considerably more information on the use of ear prints – much of it highly controversial. Scientists can seldom resist burdening the English language with ugly and unnecessary new words so it should come as no surprise to learn that the study of ear prints is sometimes referred to as 'earol-ogy'. Ear prints are typically left at a crime scene when a person was listening at a door or window before committing the offence. For example, in May 1996, a frail elderly woman was smothered to death during a burglary in Fartown, Huddersfield. At the subsequent trial, in 1998, Mark Dallagher, known locally as a petty burglar, was charged with the woman's murder. Despite protesting his innocence, claiming that an ankle injury would have made it difficult to commit the crime, and supplying an alibi, he was found guilty. This was largely on the testimony of expert witnesses for the prosecution who stated that his ears yielded prints exactly the same as those found on a windowpane at the crime scene. It did not help that his alibi said that she was asleep and 'on med-ication' at the time of the offence. Furthermore, a fellow prisoner claimed that

Mark Dallagher had confessed to him while he was being held awaiting trial. However, at a retrial that ended in January 2004, he was completely exonerated when it transpired that DNA isolated from the ear prints at the crime scene did not belong to him and there were serious concerns about the reliability of the identification of the prints themselves. For example, when he first compared Mark Dallagher's ear prints with those found on the windowpane, one of the expert witnesses had written that the two were 'definitely not' the same (www.forensic-evidence.com/site/ID/DNAdisputesEarlID.html). The prison 'confession' was also dismissed as being unreliable. The use of prison informants is always controversial as prisoners often attempt to gain favours or remission by telling the authorities tales they wish to hear. Alternatively, the informant may be pressured into concocting a story or be simply dishonest or malicious.

The use of ear prints for identification is difficult because when a person presses his or her ear against a substrate it will be deformed to an extent that depends upon the pressure exerted and the substrate. Consequently, even prints from the same person will be, to an extent, variable. Furthermore, there are insufficient studies or databases to know whether or not an ear print is truly a unique identifying characteristic of an individual. Despite this, work has begun using computerized systems for analysing ears and ear prints (www.le.ac.uk/press/press/earprint.html). In addition, some police forces take ear prints as well as fingerprints and these are entered onto a central computer database. If successful, one possibility of this approach would be to identify a person from closed circuit TV recordings (CCTV) by observing the characteristics of their ears. If the ear has been pierced, then this provides a further identifying characteristic – although it may also result in the formation of only a partial ear print (Swift and Rutty, 2003; Abbas and Rutty, 2005). Ear piercings can also be helpful features to note when identifying a dead body. Even if the ear jewellery has been lost or stolen, scar formation may remain where the ear was pierced. The inexplicable popularity among young persons for piercing other bits of their anatomy with metal rings and studs also provides identifying features should they indulge in criminal acts or their body be found under suspicious circumstances.

Retinal and iris scans

Retinal and iris scanning are being used increasingly for personal identification, e.g. for high security access at airports and in military establishments. Retinal scanning involves mapping the distribution of blood vessels on the retina by shining a low-intensity infrared light through the eye and picking up the reflected light on a video camera. The blood vessels in the retina absorb more of the inrared light and therefore stand out from the surrounding tissues. The iris is said to contain over 400 distinguishable characteristics although only a proportion of these are used in current scanning technologies. Apart from overall form and colour, these characters are formed by random events during the tissues' formation and growth. The distribution of these characters is determined using algorithms and it has been estimated that the chances of two persons sharing more than 70 per cent of their iris characteristics is approximately one in 7 billion (Daugman and Downing, 2001; Daugman, 2004). Even identical twins have different iris characteristics. Iris scanning technologies are currently faster than those for retinal scanning and have the added advantage of permitting the user to be up to a metre away from the camera. There appear to be no published studies on how retinal and iris scan characteristics change after death but eye tissues usually degrade quickly and birds often eat them. Consequently, neither of these techniques is likely to be much help where there is an unidentified dead body; it would, however, be interesting to determine whether they could provide an indication of time since death. By contrast, fingerprints can be taken after death so long as the skin remains intact: even if the epidermal layer sloughs off from the dermis (a common occurrence in bodies found in water) it is possible to insert one's own finger inside the cast skin and make a print.

Tattoos

Men, and to a lesser extent women, have adorned themselves with tattoos for thousands of years. However, in modern Western societies, their use until recently has tended to be restricted to certain male groups or professions, such as sailors (Sperry, 1991). Nowadays, men and women of all ages and degrees of affluence wear tattoos. Until the skin is lost during the decay process, these tattoos remain visible and provide identifying characteristics (Figure 3.4). Should they become obscured during the decomposition process, they can be

Figure 3.4 Tattoos can be found in the most unlikely places. The more unusual the tattoo, the more useful it can prove in identification. (Reproduced from Shepherd, 2003, with permission from Arnold)

rendered visible by treatment with 3 per cent hydrogen peroxide (Haglund and Sperry, 1993). Even if the tattoos are irretrievably lost, their former presence can be predicted from finding the pigments within nearby lymph nodes (Hellerich, 1992). Although they are normally found on the outer body surfaces, some people are tattooed inside the lips or on the gums. Tattoos can indicate a relationship, lifestyle preference, mental state, service in the armed forces or gang membership. For example, there is a probability that a man with a tattoo depicting foxes racing down his spine and disappearing into his anus may be a homosexual (and given to extrovert behaviour) whilst a woman with four dots on each finger may be a lesbian (www.sccja.org/csr-tattoo.htm). Drug addicts' tattoos often include prophesies of doom such as 'born to lose', whilst those of female prostitutes are more light hearted, such as 'to hell with housework' and 'pay as you enter' inscribed in an appropriate location. Names are not a good indication of a relationship because the tattoo may remain long after the object of affection has become a distant memory. Asian gang members, especially the Japanese Yakuza, often have extensive and elaborate tattoos. However, owing to the current popularity, it is important to keep an open mind because the tattoos could represent a person's fantasy about himself rather than the reality of his existence or be the consequences of a 'good idea' at a time of excessive inebriation.

Hair

Characteristics

Our body is normally covered with hair except for the surfaces of the palms, the palmar (lower) surface of the fingers, soles of the feet and the lower surface of the toes. Although hair comes in various lengths and thickness depending upon where it grows, its basic structure remains the same. A hair consists of two portions: a shaft projecting from the skin and a root that is embedded within the skin. The hair shaft is formed from three layers of dead cells that contain large amounts of the protein keratin and are firmly bound together by extracellular proteins. Keratin is a robust molecule that is resistant to decay and there are few animals capable of digesting it. Consequently, hair remains long after the soft body tissues have decomposed. The outermost layer of a hair is called the cuticle; this region is the most heavily keratinized and it has a scale-like appearance. The middle layer contains pigment granules in the case of darkly coloured hair but these are reduced in number or absent in grey or white hair. The inner layer, which may be absent in thin hair, also contains pigment granules. The hair root extends down into the dermis of the skin where it is surrounded by the hair follicle and at the base of this is the hair bulb. Within the hair bulb are living germinal cells that are responsible for the continued growth of an existing hair or its replacement should the hair fall out. A hair with its follicle attached therefore provides a good source of DNA.

All the hairs on our body are constantly growing and then being lost and replaced. Loss may be passive through it falling out naturally or through gentle pulling – as when we comb our hair. Clothing, pillows and combs are therefore good places to look for hairs. Hairs may be transferred between people during vigorous bodily contact, such as sex, and the more vigorous the encounter, the more likelihood that transfer will take place. In cases of sexual assault, one may therefore find scalp, facial and pubic hair from the attacker on the body of the victim. In the latter situation, and also during fights, it is also common for both victim and the attacker to pull out one another's hair. In the case of an assault or homicide it is therefore important to examine the body and clothing of the victim and also the locality for hair that is different from that of the victim. Microscopic analysis of scalp hair can provide some identifying features. For example, a transverse section through a hair shaft can indicate its shape: straight hair appears round, curly hair is kidney-shaped and wavy hair is oval. Its pigmentation can also provide an indication of coloration. However, coloration can be altered through the use of hair dyes, a skilled hairdresser can soon alter the length and appearance of a person's hair and a criminal could wear a wig during or after committing a crime. The physical appearance of hair recovered at a crime scene might therefore provide some evidence but DNA extracted from the hair provides a more reliable means of identification.

Assimilation of drugs, poisons and explosives

Drugs such as methadone and poisons such as lead and arsenic can be sequestered in hair and detected long after the last dose was administered and even after death. By contrast, most drugs and poisons are eliminated rapidly from the body when we are alive and they are lost with the soft tissues during the decay process. Hair analysis is therefore valuable in determining whether a person was taking drugs either voluntarily or unsuspectingly. The latter scenario is typical in cases of 'date rape' in which the victim is rendered incapable by their drink being laced with a hypnotic such as Rohypnol® (flunitrezapam) or zolpidem: these drugs have the added problems of causing memory impairment (Villain et al., 2004). The breakdown product of Rohypnol, 7-aminoflunitrezepam, is easier to detect in hair than flunitrezepam itself because it is more basic and therefore binds better to the melanin in hair and can be detected up to a month after ingesting a single dose of Rohypnol (Negrusz et al., 2001). However, it is worth remembering that a negative result does not mean that Rohypnol was not taken because there are big differences between individuals, which makes it impossible to determine how much Rohypnol was ingested or when. Similarly, some persons use Rohypnol as a recreational drug and therefore its presence in their hair may not be an indication that they unintentionally drank a spiked drink. Coloration is an important factor in the effectiveness with which hairs assimilate chemicals. For example, in experimental rats, white hairs assimilate less methadone than black hairs (Green and Wilson, 1996). Because hair grows at a fairly constant rate (depending upon the area of the body and our age), the amount of the chemical along the length of a hair can indicate when it was administered. For example, analysis of a few strands of hair taken from the body of King George III has shed new light on the possible reasons for his fits of madness (Cox *et al.*, 2005). King George III died in 1820 and at this time it was a popular practice to keep lockets of hair from famous people. In this case, the hair was placed in an envelope that ultimately found its way to the London Science Museum. When analysed, in 2004, the hair was found to contain arsenic at levels over 300 times the level at which it causes a toxic effect. Since the 1970s it has commonly been believed that King George's madness was a consequence of him suffering from the medical condition porphyria. Porphyria is the term used to cover at least seven related blood disorders that arise through the deficiency of enzymes involved in the production of the haem part of haemoglobin. This leads to the accumulation of chemicals called porphyrins and their precursors in the blood stream. Most sufferers do not experience severe illness but in some the symptoms include seizures, anxiety and mental confusion. Effects such as these are much more common in women than in men but as arsenic is capable of triggering acute attacks of porphyria this could explain why King George suffered so unexpectedly badly. It does, however, beg the question of how he came to ingest such remarkably high quantities of arsenic. The even distribution of arsenic along the length of the hairs indicated that it was not surface contamination or the result of a single poisoning but had been acquired throughout the time the hair was growing. King George's medical records mention the use of skin creams containing arsenic but this would have been insufficient to explain the amounts found in his hair. Similarly, the common practice of wearing wigs that were powdered with dust containing arsenic is also thought to be an unlikely explanation – although neither of these would have helped. The answer probably lies in him being treated with potions such as emetic tartar – antimony potassium tartrate – to alleviate his mental confusion. At the time, doctors believed passionately in the powers of bleeding, purging and vomiting for the treatment of virtually all ailments and, as its name suggests, emetic tartar was used to induce vomiting. Unfortunately, antimony and arsenic often occur together in minerals and it was then impossible to completely separate the two. Consequently, in an attempt to cure King George of his 'madness', his doctors were actually making his condition worse.

A study of Napoleon Bonaparte's hair has revealed that he too suffered from arsenic poisoning. Napoleon's death has been a source of controversy from the moment he died on the isolated island of St Helena in 1821. An autopsy conducted at the time indicated that he died of stomach cancer – the same disease that killed his father. However, recent analyses of his hair have indicated levels of arsenic '7–38 times above normal' (http://news.bbc.co.uk/1/hi/world/europe/1364994.stm). How the arsenic was ingested has spawned a mix of hypotheses ranging from the banal to the criminal. Some workers have suggested that the arsenic may have come from wallpaper or paint in the house where he lived whilst others think that he was assassinated by the British to stop him escaping and raising another army. Another possibility is that, like King George III, he became the unintentional victim of his doctors. To alleviate his stomach pains, they also prescribed him frequent doses of antimony potassium tartrate. In addition, he received regular enemas and purges that would have further weakened him. It is possible that Napoleon would have died anyway from stomach cancer but his doctors probably helped speed him on his way.

Interestingly, it is possible to detect traces of explosives in hair, even after it has been washed (Oxley *et al.*, 2005). This includes explosives such as TATP (triacetone triperoxide – also known as acetone peroxide and 'Mother of Satan'), which was used in the London Tube Train bombings in July 2005. Many explosives have low vapour pressures but even these may become either adsorbed or absorbed, or both, onto hair. However, traces of explosives may also be assimilated onto hair by contamination from hands or sharing a contaminated towel. Although there appears to be no published information on the subject, it is possible that traces may also be transferred by head to head contact during kissing – and this is a very common form of greeting between men in some societies. Consequently, detecting traces of explosives on a person's hair is not necessarily proof that they were involved in terrorist activities.

Bones

Characteristics

The skeleton of an adult human is comprised of 206 named bones but there are even more in a child. The reason for this discrepancy is that some of our bones, such as the hipbones, fuse together as we get older. The skeleton of an adult can be divided into two compartments: the axial skeleton and the appendicular skeleton. The axial skeleton comprises the skull, hyoid bone, the bones of the ears, the vertebral column and the thorax. The appendicular skeleton is comprised of the bones of the shoulder blades, the upper and lower limbs and the hip girdle. Most of the bones are paired on the left- and right-hand sides of the body. The bones can be classified into five types based upon their shape: long, short, flat, irregular and sesamoid. The long bones are those such as the femur in the thigh. These bones are longer than they are broad and are usually slightly curved because this shape allows weight to be more evenly distributed along their length. Short bones are approximately as broad as they are long and typical examples are the majority of the carpal bones found in the wrist. As their name suggests, the flat bones are flat and 'plate-like'. They consist of two almost parallel layers of compact bone and examples include the bones of the skull, the breastbone and the ribs. These bones provide protection for the underlying soft tissues and an extensive area for the attachment of muscles. The irregular bones are those of the vertebral column and some of the facial bones. The sesamoid

Figure 3.5 It can be easy to mistake both natural and man-made objects for bones. This piece of wood bears a superficial resemblance to a large femur

bones gain their name from their shape, resembling that of a sesame seed. They are found within those tendons that bear heavy stresses and strains and they protect them from excessive wear. Most sesamoid bones are only a few millimetres in size, such as those in the palms of the hand, and their number can vary between individuals. An exception to this generalization is the two patellae or kneecaps. In addition to these bones, there are the small sutural bones that are found between the sutures of some of the cranial bones. The number of these sutural bones varies considerably between individuals.

The identification of a complete human skull or the larger bones is relatively easy but distinguishing the small bones and sesamoid bones from those of other animals requires more skill. This can be important where skeletons have become disarticulated and only partial remains can be found. Consequently, it is not unusual for the police to be alerted to the presence of bones on a beach or moor that subsequently prove to be those of a sheep or pig. Even oddly shaped pieces of wood and stones are occasionally confused with bones (Figure 3.5). The anatomical measurement of bones can provide an indication not only of whether or not they are of human origin but also of gender, ethnicity and age at the time of death. The bones may also provide clues to how long a person has been dead, the presence of any underlying diseases, the cause of death, past movements and, if DNA is extracted, identity (Byers and Myster, 2005).

Determination of gender

Gender determination is seldom a problem whilst a body is still in the early stages of decay because the genitalia and secondary sexual characteristics (e.g. facial hair) are present. Individuals who have undergone a sex change operation can present more difficulties but an examination of the internal reproductive organs usually solves the matter: the uterus and the prostate glands are amongst the last soft tissues to decay. If all that remains is the skeleton, it is still possible to determine the gender with a high degree of accuracy (Duric *et al.*, 2005). The most helpful indicator is the pelvis, which exhibits numerous differences between men and women. For example, in a woman the pelvis is smoother, lighter and more spacious than in a man. Similarly, the area of the pelvis called the sub pubic arch has an inverted 'U shape' in women but an inverted 'V shape' in men and there are several other differences. There are also gender differences in the dimensions of the bones of the skull (especially the mandibles), the long bones and the sternum. Some of these bones cannot by themselves provide an accurate determination of gender but when used in combination with two or more other bones the accuracy can be increased to more than 95 per cent.

Determination of ethnic origin

Ethnic origin is extremely difficult to determine solely from skeletal remains and the skull is the only structure of use. Although it is possible to assign a skull to one of three broad groupings, i.e. Caucasian, Negro and Mongoloid, further discrimination is seldom possible. For example, one cannot distinguish the skull of a White Englishman from that of a White Austrian or Estonian or that of a Black Englishman from a Black Nigerian or Mauritanian. This is partly, of course, because there is no such thing as a 'pure race' and many anthropologists, such as Brace (1995), consider that it would be more helpful to consider human variation in terms of clines of traits rather than discrete populations. Anatomical differences that have proved useful for distinguishing between races include the size of the nasal openings – these tend to be narrower and higher in Caucasians than Negroes – and the structure of the mandibles. Krogman and Iscan (1986) have disputed the value of mandible measurements for racial discrimination but Buck and Vidarsdottir (2004) were able to use a computer-based method of geometric morphometric analysis of mandibles to identify unknown sub-adult individuals with an accuracy of over 70 per cent. This is noteworthy because up until a person reaches adulthood many anatomical characteristics are constantly changing in proportion to one another owing to the fact that growth does not occur uniformly. Consequently, most of the work on race has been done on adult skeletons.

Determination of stature

Stature refers to a person's natural height when standing in an upright position and, clearly, it is an important consideration in the identification process.

Although it would appear to be a simple measurement to make, it presents dif-ficulties even whilst a person is alive – once they are dead, things only get worse (Bidmos, 2005). Stature varies naturally throughout the day owing to differen-tial loading on the spinal vertebrae and errors are commonly made when making the measurements. If the only records available are those made by the individ-ual concerned, they should be treated with some suspicion. Men, especially, tend to overestimate their height. Furthermore, as we enter old age, our bones tend to shrink and our posture changes. Consequently, during adulthood, stature is not a constant but changes gradually with time. Once a body has become skele-tonized, determining stature is not simply a matter of laying the bones out on a bench and measuring the length of those that contribute to height. This would not be the way the bones would be arranged in relation to one another during life and the cartilage at the joints would be missing. A further problem that can arise is that some of the bones might be missing. Nevertheless, provided that one has at least one of the long bones – such as the femur, tibia or humerus – it is possible to estimate stature with reasonable accuracy. The length of the long bones is proportional to height, so by reference to a table or regression equa-tion one can arrive at an estimated height (Trotter, 1970). Obviously, the accu-racy of this approach increases with the number of long bones available for measurement and allowances have to be made for gender and race. There can also be significant individual variation – for example, we are all familiar with people who have unusually long or short legs in proportion to the rest of their body.

Determination of age

If all the soft body parts are decayed, one could use a combination of both dental and skeletal characteristics to estimate how old a person was when they died. These characteristics are not especially accurate and become less so as a person ages. In terms of skeletal development, remains can usually be catego-rized into one of eight groups: perinatal (foetal), neonatal, infant, young child-hood, late childhood, adolescence, young adult and older adult. The term 'perinatal' refers to unborn babies and their age can be determined with some accuracy from measurement of the cervical, thoracic and lumbar vertebrae (Kosa and Castellana, 2005). The development of an unborn child proceeds at a regular and predictable rate because the foetus is protected from the outside environment and if food is short its growth will be at the expense of the mother. Neonates are babies that have been born but have not yet developed any teeth. At this time, the baby has very small bones and many of these, such as the pelvis and those of the skull, are not yet fused together. However, there is a lot of indi-vidual variation in the speed with which these events take place. In infants and young children, the teeth begin to erupt and these provide a fairly accurate indication of age. In young children, the bones start to ossify (harden) although, again, there is a great deal of variation between individuals in when and how rapidly this process develops. In late childhood, more of the bones begin to ossify and the permanent teeth make their appearance.

During adolescence, the long bones grow rapidly in length. This is brought about by the activity of the chondrocyte cells in the regions of the bones called the epiphyseal plates. The chondrocytes multiply and form a layer of cartilage that causes the epiphyseal plate to become wider – hence the bone elongates. As the cartilage is formed, the chondrocytes on either side of it die off and their place is taken by another cell type, the osteoblasts. The osteoblasts then convert the cartilage into bone and so the bone shaft grows. By late adolescence, the cartilage of the epiphyseal plates becomes completely replaced with bone tissue in a process known as epiphyseal plate closure. Once closure is complete, further lengthening of the bone is not possible although changes may still take place in circumference. The number of epiphyseal plates differs between bones and the timing of their closure varies both within and between the different types of bones. It is therefore possible to estimate age with reasonable accuracy on the basis of the extent of epiphyseal plate closure within the skeleton. For example, in the clavicles (collarbones) of men, epiphyseal closure at the acromial end of the bone (i.e. that next to the scapula) occurs at 19–20 years of age. However, at the opposite sternal end (i.e. that next to the breastbone), epiphyseal closure does not occur until 25–28 years of age (Krogman and Iscan, 1986). The levels of the sex hormones also heavily influence the timing of plate closure during puberty. Consequently, gender has to be considered when attempting to estimate age from skeletal characteristics.

After adolescence, age determination from skeletal remains becomes more problematic. One method is to observe the degree of closure of the cranial sutures in the skull – as the plates fuse together with time, the sutures become less distinct. However, this is not very accurate. An alternative approach is to observe the shape of the rib bones and the degree of pitting of the cartilage that connects the ribs to the breastbone (Iscan et al., 1984; Kunos et al., 1999). To begin with, the ends of the ribs are flat and the cartilage is smooth, but with increasing age the rib ends become ragged and the cartilage becomes pitted. Unfortunately, the reliability of this approach has been questioned (Schmitt and Murail, 2004) and the rib shafts themselves are fragile and, along with the cartilage, tend to decay fairly rapidly if the corpse is left in an exposed position or an acid soil. Another method is to look for evidence of age-related degenerative conditions, such as arthritis or osteoporosis. However, many of these conditions afflict people at varying times and even children can suffer from arthritis (juvenile idiopathic arthritis is different from adult arthritis but can still result in damage to the joints). Evidence of disease conditions affecting bones can, however, prove useful when attempting to establish the identity of skeletal remains.

Facial reconstruction

Facial reconstruction from skull features is normally done manually by the application of modelling clay to a cast of the skull, although computer-aided techniques are becoming more sophisticated, have the added advantage of being quicker and cheaper and the final images can be transmitted more easily between interested parties (Vanezis *et al.*, 2000; Wilkinson, 2004; De Greef and Willems, 2005). There are two basic approaches to the reconstruction process: the anatomical method and the tissue depth method (Kahler *et al.*, 2003). The former method requires a great deal of anatomical knowledge because the skin and all the underlying tissue layers must be built up gradually layer by layer, starting from the skull surface. Consequently, these days, the anatomical method is usually restricted to the reconstruction of fossil humans and hominids for which there are no statistical data on facial features. The tissue depth method uses statistical data banks of average tissue depths between the bone and skin surface at various marker points on the skull. The tissue depths between the marker points are then interpolated and finally a 'skin surface' is applied. Obviously, the tissue depths at the marker points are affected, amongst other things, by build, gender, age and racial characteristics but extensive databanks are now available for several racial groups. Unfortunately, it is not possible to predict the shape of the mouth, the nose or the eyes, and unless hair has been preserved there will be no indication of hair colour or length. As mentioned previously, it can also be difficult to predict racial origin, and hence skin colour, from skeletal remains. Consequently, two reconstruction experts working independently from the same skeletal remains may produce rather different models. Indeed, owing to the limitations of existing techniques, it has been suggested by Stephan and Henneberg (2001) that facial reconstructions be limited to situations in which all other methods of identification have failed.

Determination of age of remains

Determining the age of skeletal remains that are less than 100 years old is difficult because there are few reliable morphological changes during this time and current chemical tests are relatively imprecise or still in the experimental stages. For example, the association of ligaments and other soft tissues with the bones is sometimes stated to indicate that the remains are less than 5 years old whilst the presence of blood pigments within the bones indicates that they are less than 10 years old. However, these features will be strongly influenced by the environment in which the body has been exposed. Another approach to estimating the age of human bones is to analyse their isotope levels and ratios. The vast majority of elements present in our bodies and the environment exist as a variety of isotopes, i.e. their nuclei contain the same number of protons but a different number of neutrons. Some of these isotopes are unstable and decay through the loss of sub-atomic particles into other isotopic forms. For example, carbon-14 decays to nitrogen-14 through the loss of beta particles. The rate of decay of an element, measured as the 'half-life', depends on the instability of the isotopic form and may vary from fractions of a second to millions of years. The levels of unstable isotopes, especially carbon-14, have proved useful in determining the age of both human and animal remains. Carbon-14 has a half-life of 5730 years and would have disappeared from the Earth millions of years ago except for the fact that it is constantly being formed by the interaction of nitrogen in the air with cosmic rays. Once formed, the carbon-14 becomes incorporated into carbon dioxide, which plants then metabolize into organic molecules when performing photosynthesis. The carbon-14 subsequently becomes incorporated into the bodies of herbivores when they eat the plants and then into the bodies of any carnivores or parasites that feed on the herbivores – consequently there is a constant cycling of carbon-14 between all living organisms. However, once an organism dies it no longer acquires new carbon-14 and the level within its body slowly declines. Because the rate of radioactive decay is constant, the level of carbon-14 present in the body provides an indication of how long the organism has been dead. Obviously, allowances have to be made through calibrations for the effects of complicating factors that influence the level of carbon-14 in the atmosphere, such as solar storms and human activities (e.g. the burning of coal and other fossil fuels). Carbon-14 dates are usually expressed as 'years before present' (BP), present being the year 1950 – the year in which extensive above-ground nuclear weapons testing began and thereby dramatically increased the levels of carbon-14 in the atmosphere. Carbon-14 dates are always cited as plus and minus a standard deviation (e.g. 526 ± 40 BP) to allow for errors that arise through the nature of the sample, how it was collected and the methodology used. It is therefore impossible to state the exact year in which a person died. Furthermore, current techniques for carbon-14 dating are also insufficiently precise to accurately age the skeleton of a person who died within the last 100–200 years. However, if the remains are older than this the police do not need to proceed further: they do not investigate crimes that were committed more than 70–75 years ago because the perpetrator would almost certainly have died and the victim's family would be several generations removed. Despite the apparent limitations of carbon-14 dating, Spalding *et al.* (2005a) have developed a method for measuring carbon-14 levels within DNA that allows them to estimate the ages of cells to within 2 years. It would therefore be interesting to see whether their technique could be utilized for forensic purposes.

Carbon-14 is not the only isotope that can be used to determine the age of human remains. For example, our bones contain small quantities of lead-210 that we acquire through our food and breathing in radon-222 (this decays into lead-210) from the atmosphere. Because lead-210 has a half-life of only 22.3 years its levels decline more rapidly after death than those of carbon-14, thereby making it potentially valuable for forensic studies (Swift *et al.*, 2001). Isotopes that do not undergo radioactive decay are said to be 'stable'. An element may posses both unstable and stable isotopes, e.g. carbon may occur as both the unstable isotope carbon-14 and the stable isotopes carbon-12 and carbon-13. Stable isotopes have many uses in forensic studies because their ratios can prove a unique identifying feature of a specimen's provenance or as an indication of previous diet or geographical origin (Anon, 2004a). For example, strontium posses four stable isotopes (strontium-84, -86, -87 and -88) and because their ratios vary between geographical locations these can act as a 'signature' indicating where a person or animal was living. Strontium enters our bodies via our diet and becomes sequestered in the bones and teeth. Once formed, the tissues that make up teeth have a low metabolic turnover and the strontium is immobilized within them. The strontium isotope ratio of the teeth therefore tends to reflect that of the environment in which a person grew up. By contrast, bones are metabolically more active, with new tissues being formed constantly throughout life, although the rate of renewal varies between bones. The strontium isotope ratios found in bones therefore reflect those in the environment where a person has lived in the previous 10 years. Beard and Johnson (2000) analysed strontium isotope ratios to differentiate between the disarticulated skeletal remains of three American servicemen that were found mixed together in a shallow grave in Vietnam 20 years after the conflict ended. The identity of the victims was known but the remains needed to be separated from one another so that they could be returned to the appropriate family. The three servicemen grew up in different areas of America so it was possible to identify at least some of the remains by comparing the strontium isotope ratios of the teeth and bones with those found in the three regions. Because the servicemen were not stationed in Vietnam for long before they were killed, the strontium levels in the bones were not thought to have been affected by the food and water they consumed whilst there. Although this type of analysis offers promise for future development, its effectiveness is ultimately dependent upon the availability of databases of isotope ratio analyses for different geographical regions and is potentially subject to many complicating factors. For example, human populations are increasingly mobile and our food and water are acquired from different geographical regions to where we live.

Teeth

Characteristics

The forensic importance of teeth arises from them containing a great deal of personal information coupled with being the most indestructible part of the body. Adults normally have 32 teeth: on each side of the mouth there are 4 incisors, 2 canines, 4 premolars and 6 molars. Each tooth has a characteristic morphology but they all consist of three regions: the portion above the gum line is called the crown, at the gum line is a constricted region called the neck, whilst the portion embedded beneath this within a socket in the jaw is called the root. The canines, incisors and first lower premolars each have a single root whilst the first upper premolars tend to have two roots. The first two molar teeth tend to have two roots whilst the corresponding upper molars tend to have three roots. The number of roots attached to the third molar teeth can vary, although most have a single fused root. Teeth owe their indestructibility to being composed largely of dentine – this is a highly calcified connective tissue that is responsible for giving the teeth their shape and rigidity. Calcium salts comprise approximately 70 per cent of the dry weight of dentine, thereby making it harder than bone. In the crown region, the dentine is covered with a layer of enamel: calcium salts make up about 95 per cent of the dry weight of this region so it is even harder than dentine. Within the root region, the dentine is covered with a thin bone-like layer called the cementum. At the centre of a tooth lies the pulp cavity, which contains connective tissue, blood vessels and nerve endings.

Determination of age

We have two sets of teeth, or dentitions, during our life: the deciduous teeth (also known as the 'milk teeth', 'primary teeth' and 'baby teeth') and the permanent teeth. The deciduous teeth begin to emerge from the gums about 6 months after birth. Thereafter, at about monthly intervals, a further pair emerges until all 20 deciduous teeth are present. The deciduous teeth start to be lost when we reach about 6 years of age and they have usually all gone by the age of 12. They are replaced by the permanent teeth – which include some additional molars. The wisdom teeth usually do not emerge, if they emerge at all, until after the age of 17. It is therefore possible to age a child with a reasonable degree of accuracy on the basis of the teeth that are present and their stage of development.

Our teeth start to form even before we are born and by the time a foetus is 4 months old mineralization has already begun. Birth is a physiologically traumatic event for the baby and one of its consequences is to upset the production of dental enamel. The enamel is laid down as a series of lines called the striae of Retzius and at birth a 'neonatal line' is formed that is darker and bigger than the surrounding striae – this can be seen when a tooth is sectioned (Skinner and Dupras, 1993). The presence of a neonatal line within the first deciduous teeth or at the tips of first permanent molars therefore indicates that a child survived birth and lived for at least a short time afterwards. This is therefore an important observation when the body of a baby is discovered.

Although it is possible to use dental characteristics to age children with a reasonable degree of accuracy, it is much more difficult for adults. One of the most promising methods of overcoming this problem is to compare the ratio of left

(L-) and right (D-) isomers of the amino acid aspartic acid within the dentine (Ohtani *et al.*, 2003). We, like all living organisms, utilize only L-amino acids and when dentine is first formed it contains only L-aspartic acid. However, once the dentine is formed, the aspartic acid within it undergoes a process called racemization in which the molecules rotate until there is a 50:50 mix of L- and D-aspartic acid. Because the rate of racemization is known, by measuring the ratio of L- and D-aspartic acid it is possible to estimate a person's age to within about a year. An older, but more established approach relies on sectioning a tooth longitudinally and measuring the translucency within the dentine. As a tooth ages, the dentine becomes more mineralized and this causes it to change in appearance from being opaque to increasingly translucent. Unfortunately, this technique is only accurate to about plus or minus 6 years.

Spalding *et al.* (2005b) have described a novel method of age determination based on measurements of carbon-14 levels within teeth. Their technique utilizes the absence of carbon turnover within the enamel once it is formed. This means that the carbon-14 level within the enamel reflects the level present in the atmosphere when the teeth were being formed. The atmospheric levels of carbon-14 are well documented: there was a rapid rise in the atmospheric levels of carbon-14 when above-ground nuclear testing began, followed by an exponential decline once the tests were banned. The age at which the various teeth develop is also well known, therefore the carbon-14 levels indicate the year in which the enamel was produced whilst the tooth type indicates the age of the person when this occurred. For example, the wisdom teeth are the last ones to be formed, their enamel being laid down when a person is about 12 years old. If the carbon-14 levels of a person's wisdom teeth correlated with the atmospheric levels in 1974, then the person was probably born in 1962. Analysing carbon-14 levels in a variety of teeth improves the accuracy of the technique. This method still needs to be verified and the effect of factors such as diet assessed, but it offers good prospects for determining the age of persons who were children at, or born after, the start of nuclear testing.

Determination of gender and ethnic origin

It is difficult to determine a person's gender from the physical appearance of the teeth and in any case this would normally be done from other morphological evidence. However, if only the teeth are available, it is possible to extract DNA from the pulp and test for Y chromosome-specific repeat sequences (Murakami *et al.*, 2000). Although there is some variation in tooth morphology between ethnic groups, it is very difficult to come to any firm conclusions based solely on these characters (Edgar, 2005). For example, persons of Asian and Native American ancestry tend to have shovel-shaped incisors whilst Europeans often demonstrate specific features on their deciduous posterior premolars and permanent molars that are collectively referred to as Carabelli's trait. However, even these features are highly variable between populations: Correia and Pina (2000) recorded Carabelli's trait in 85 per cent of White North Americans but only in 13.5 per cent of Portuguese. The nature of any dental work can, however, provide an indication of a person's origins. Typically, the materials used to prepare dentures and bridges and the quality of the workmanship can indicate when and where the work was done.

Identification based on dental characteristics

By noting the presence or absence of teeth, their appearance and the nature of any past dental work it is possible to build up a person's dental profile. This procedure is known as 'dental charting' and until the advent of DNA technology it provided one of the most accurate means for confirming a person's identity (Pretty and Sweet, 2001; Adams, 2003). It relies on comparing the chart of, for example, an unidentified dead body with those held on dental or medical records. Unfortunately, within the UK, there is no legal requirement for dental practitioners to retain dental records for a specific period. Further problems can arise if there were errors in dental charting pre- or postmortem, if the dead person had not visited a dentist for a long time, was a recent immigrant (and therefore had no records in this country) or the body was destroyed in some way such that only a partial collection of teeth was left for study (Whittaker, 1995). Consequently, even if records are available, there may not be an exact match between the dental chart of the dead body and that of the most likely candidate. The situation is alleviated if there are premortem radiographic records that can be compared to those taken from the dead body. If the radiographs match then it is possible to confirm identity with a high degree of certainty. However, in the absence of radiographs, at this point, the forensic dentist (also called a forensic odontologist) has to make a value judgement to decide if there is a rational explanation for any discrepancies. Clearly, the presence of a permanent tooth that the premortem chart stated had been extracted indicates that the candidate must be excluded. However, the absence of a tooth that the premortem chart states should be present could be explained by it being lost in an accident or a fight or extracted by another dentist subsequent to the records being made – it is now common practice for persons in the UK to go abroad for dental treatments. Although it is obviously preferable to have numerous points of concordance between premortem and postmortem dental charts, there is no agreed minimum number that are required to confirm a positive identification. Recently, a computer program called OdontoSearch has been developed (www.cilhi.army.mil) that can facilitate chart comparisons – however, unlike NAFIS or IAFIS, it is not a database from which a specific individual can be identified.

Despite the potential problems, dental charting remains a valuable method for identification. It is especially important when there are large-scale disasters such as plane crashes or explosions that produce numerous badly damaged unidentified dead bodies. For example, dental examinations were a central feature of the identification of the victims of the 'Bali bombing' (www.defence.gov.au/dpe/dhs/infocentre/publications/journals/NoIDs/adfhealth_ sep03/ADFHealth_4_2_50-55.html). Bali is an island belonging to Indonesia that is a popular tourist destination. At about midnight on12 October 2002, a huge bomb was exploded outside the Sari Club in the town of Kuta. The radical Islamic group Jemaah Islamiah, with the intention of killing Americans, planted the bombs. However, far more Australians and Indonesians were killed than Americans. The bomb was poorly manufactured but it still caused enormous devastation and a fierce fire followed. A total of 202 people are believed to have died and many of the bodies were torn apart by the explosion or cremated in the fire that followed. This made visual identification from appearance, clothes or possessions impossible in some cases. Many of the Australian victims were young (less than 30 years old) and, like young people in the UK, they had been brought up with fluorinated drinking water and a good dental service. Consequently, they often had little evidence of dental work that could be used to aid the identification process. However, the presence of dental braces and crooked, chipped or spaced teeth could still provide a characteristic pattern that could be compared to charts and photographs in which the victim was exposing his teeth whilst smiling.

Effects of drug abuse on dental characteristics

Poor dental health, especially in young adults, may be a consequence of underlying disease or neglect but could also raise suspicions of drug abuse. Heroin addicts are notorious for possessing bad teeth, especially if the drug is taken through 'tooting straws'. Similarly, methadone is supplied as sugary syrup that can encourage tooth decay. Cocaine addicts usually test the quality of their drug by placing a sample on their teeth. This can cause the surrounding gums to shrink and then bleed spontaneously. LSD, Angel Dust and other hallucinogens can induce teeth grinding resulting in excessive wear – however, this is also a common nervous symptom and should not be taken as evidence of drug use. Solvent abusers often develop a characteristic 'glue-sniffers rash' around their mouth and nose and the tips of their nose may exhibit signs of frostbite as a consequence of inhaling aerosols. Drug users and drunks often lose and damage teeth as a consequence of falling or getting into fights whilst 'under the influence'. However, it is also a feature of persons indulging in physically dangerous contact sports such as rugby and boxing. Dental characteristics should therefore be evaluated in the context of other forensic evidence.

For a summary of the forensic evidence that can be obtained from human tissues, see Table 3.1.

Table 3.1 Summary of forensic evidence that can be obtained from human tissues

Tissue	Forensic evidence	Test
Skin	Identification	Fingerprints DNA profiling Tattoos Physical characteristics Scars
	Cause of death / sequence of events during an assault	Wound analysis
	Contact with chemicals (e.g. explosives)	Specific tests
	Drug abuse	Scars at injection sites
Hair	Identification	DNA profiling Physical characteristics
	Contact with chemicals (e.g. explosives, drug use)	Specific tests
Nails	Identification of assailant	DNA profiling of material trapped under nails
Bones	Identification	DNA profiling Physical characteristics
	Time since death	Isotope analysis Physical characteristics
	Geographical origin	Isotope analysis
	Cause of death / sequence of events during an assault	Wound analysis
Teeth	Identification	DNA profiling Dental charting Physical characteristics Racemization Isotope analysis
	Geographical origin	Isotope analysis
Eyes	Identification	Retinal scanning (live persons only) Iris colour (fades after death)
	Time since death	Specific chemical tests
Internal organs	Cause of death / sequence of events during an assault	Wound analysis

Wounds

Definitions

Wounds can be crudely divided into those that are inflicted through the delivery of kinetic energy (i.e. a physical force), such as from a punch or knife thrust, and those caused by non-kinetic energy, such as burning or scalding. However, it is not unusual for a combination of both mechanisms to be involved, such as when a person is shot at close range. There are several good medical textbooks that provide extensive coverage on the examination and interpretation of wounds (e.g. Saukko and Knight, 2003; Shepherd, 2003), so only a brief coverage will be provided here. Medically, traumas (injuries) resulting from physical forces can be categorized into those caused by either blunt objects or sharp objects. Blunt force traumas include bruises, abrasions and lacerations whilst sharp force traumas are divided into incisions and penetration (stab) wounds (see Table 3.2). Under English law, the legal definition of a wound is more restrictive: it is a situation in which the whole skin must be broken (this includes the skin within the cheek or the lining of the lip) (www.cps.gov.uk/legal/section5/chapter_c.html#10). Consequently, a slight scratch would not be a wound, but

Table 3.2 Summary of wound types and their causes

Wound type	Results from	Typically caused by
Bruise	Internal blood loss from damaged blood vessel. May be superficial or deep	Heavy blow. Restraint by gripping or pressing against a hard surface
Abrasion	Superficial damage to skin surface. May or may not be accompanied by blood loss	Contact with a rough surface. Dragging of body across an object or of an object across the body. Hanging, garrotting
Laceration	Overstretching of skin resulting in tearing	Glancing blow from heavy object
Incised	Cutting of the skin and underlying tissues. Wound is longer than it is deep. Often superficial unless weapon is large, such as a sword	Object with a sharp, rigid cutting surface, such as a knife or broken bottle
Stab	Deep penetrating wound severs blood vessels and damages internal organs	Rigid object with a sharp point, such as a knife

if the scratch bled it would be. Similarly, a ruptured artery would not be a wound so long as the bleeding took place internally and not through a break in the skin. Injuries such as bruises, fractures and other internal injuries would therefore lead to prosecution under a different criminal offence. For the purposes of this section, wounds will be dealt with from a medical perspective. The trauma categories mentioned above are not mutually exclusive and a wound may exhibit more than one characteristic. For example, a bite may show both bruising and penetration of the skin surface. The nature and distribution of wounds on a body can provide a wealth of forensic information, such as when and how they were inflicted and whether they were the result of foul play, an accident, self-harm, suicide or were caused during the process of decay or even during the handling and storage of the body at the mortuary.

The formation of a wound results in the death of cells and if the victim remains alive it is followed by an inflammatory reaction in which cell debris and, if present, any fragments of the object causing the wound are removed. During this stage, immune cells (e.g. lymphocytes, macrophages and neutrophils) produce inflammatory chemicals called cytokines, such as interleukin-1α, interleukin-1β and tumour necrosis factor α. The inflammatory substances cause the blood capillaries to dilate and become more permeable, thereby allowing white blood cells, cytokines and the chemicals responsible for the blood clotting process to reach the injured region. Whilst the inflammatory process is ongoing, tissue repair commences with the stage known as 'organization' or 'proliferation' in which the blood supply is restored to the damaged region. In this stage, any blood clots are replaced with granulation tissue, which contains delicate capillaries and proliferating fibroblasts (these cells secrete collagen, elastin and other proteins that provide structural stability). Finally, there is the stage of regeneration and scar formation, although the capacity for this varies between tissue types and the nature of the wound. The sequence of these histological (cellular) and chemical changes can be used to age wounds and this can be important in cases of alleged abuse or torture and when there is a dispute about when a fight took place. For example, the suspect in a case of homicidal assault may have wounds to his knuckles but claims that these were received during a fight with another man either before or after the date on which the victim died. Ohshima (2000) has described how the levels of a number of cytokines and the messenger RNA sequences that code for them might be used to age wounds, but more work needs to be done to confirm their reliability and how levels change after death.

Bruises

Bruises (also referred to as contusions and ecchymoses) result from the escape of blood from damaged blood vessels and, in addition to those that form beneath the skin, they may also occur within muscles and internal organs. The leaked blood gives rise to the characteristic discoloration beneath the skin surfaces, although this would not be visible in deep bruises. Bruising is more obvious in persons with pale skin, such as red-heads and may be masked by skin pigmentation in Black people. A bruise may not form at the injury site because the leaking blood flows via the path of the least resistance and then collects where it can travel no further. Consequently, a broken jaw may give rise to bruising on the neck. Bruises are commonly formed from forceful contact with a blunt object – such as would occur during a fall, being gripped tightly or being hit with a fist or stick. They are seldom a feature of suicide. Although the size of the bruise often relates to the degree of violence with which the blow was inflicted, caution is always needed in their interpretation. For example, elderly people and chronic alcoholics bruise easily. The amount of subcutaneous fat, the tautness and underlying support of the skin and the nature and abundance of its blood vessels are all further factors influencing the extent to which bruises develop. Consequently, it requires considerable force to cause bruising to the palms of the hand or soles of the feet, whilst the buttocks and the regions around the eyes and genitals bruise easily. Similarly, women bruise more easily than men because they tend to have a thicker layer of subcutaneous fat, and obese persons bruise more readily than those who are slim with good muscular tone.

It is difficult to age bruises because the changes in coloration that take place during healing depend on many variables such as the size and depth of the bruise, the body part affected, gender, health and age. Despite this, the presence of multiple bruises of varying stages of healing is strong evidence of repeated assaults, such as would occur during child abuse, wife beating or prolonged torture. Even here, though, there are alternative explanations, such as underlying blood disorders (e.g. leukaemia and idiopathic thrombocytopenia purpura) and medical conditions affecting balance (and hence a propensity to fall over) can result in the sufferer acquiring regular bruises.

Recently formed bruises are dark red in colour and provide the most accurate impression of the object that caused the injury. These bruises can therefore provide most information provided that they are examined as soon as possible or if the recipient died soon afterwards and therefore the bruise had not started to become diffuse. For example, the pressure exerted via the fingers in manual strangulation and forcible restraint results in a series of bruises slightly larger than the fingerpads themselves (Figure 3.6). This can prove useful when attempting to determine the course of events that led up to death. The pattern of bruising may indicate a single event or repeated strangulation both before and at the time of death. The latter scenario can result from sex games that have gone wrong (erotic asphyxia is a dangerous practice), wife / partner beating or because the murderer has strangled his victim to unconsciousness, allowed them to revive and repeated the exercise. The bruises may be accompanied by scratch marks from the fingernails of either the assailant exerting the pressure or the victim attempting to remove the hand (Figure 3.7). Damage to blood vessels

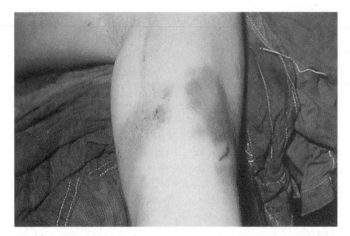

Figure 3.6 Bruising caused by forceful restraint. Note the fingerpad marks and the abrasions caused by the fingernails. (Reproduced from Shepherd, 2003, with permission from Arnold)

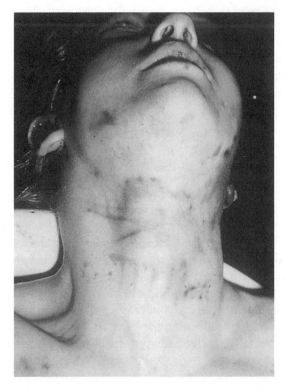

Figure 3.7 Bruising caused by manual strangulation, some of which resulted from the victim attempting to remove the murderer's hands. Note also the scatching caused by the fingernails. (Reproduced from Shepherd, 2003, with permission from Arnold)

that occurs after death does not normally result in significant bruising because the circulation has ceased and therefore blood is not forced out of the vessels. Consequently, it would be possible to distinguish between a suicidal hanging and a murderer who manually strangled or suffocated his victim and subsequently attempted to cover the crime by suspending his victim from a rope. A further distinguishing feature would be the presence of counter-pressure bruising resulting from the victim being held down forcibly on a hard surface when the initial attack took place. This can result in bruising to the back, shoulder blades or pelvis if pinned face up or to the chest, face or knees if pinned face down. It should be noted that strangulation does not always result in tell-tale surface bruising; sometimes the bruising is only apparent in the underlying neck muscles. In cases of rape, bruising might be found on the inner thighs, especially if the attack took place whilst the victim was alive and conscious, along with counter-pressure bruising as outlined above and fingerpad bruising to the arms, neck or face as result of forceful restraint.

Deep bruises to the muscles and organs normally result from extremely forceful blows or pressure and in some circumstances may prove fatal. These bruises are typically, although not always, caused by being hit with a wide blunt object, such as a kick from a heavily shod foot. Deep bruises are not always accompanied by damage to the skin surface – a lot depends on the region of the body that was hit and the object used. For example, a kick to the abdomen may result in laceration of the viscera or rupture of the spleen without causing bruising of the skin surface. Attacks involving kicking, however, often involve numerous blows being inflicted and these are typically also aimed at the genitals, face and kidneys, so bruises would be found elsewhere on the body.

Abrasions

Abrasions (scratches and grazes) are superficial injuries resulting from the body hitting or being dragged across a rough surface – or *vice versa*. They can also result from the skin being crushed and, depending on the depth, may or may not result in blood loss. Obviously, the amount of damage will reflect both the force and the nature of the surface striking the skin. If the surface is relatively smooth, it may result in such fine linear abrasions that the surface of the skin simply appears reddened – this is commonly known as a friction burn. Unlike bruises, abrasions only occur where the causative force was applied. Abrasions can indicate the direction in which the force was applied because torn fragments of skin are dragged towards the furthest edge of the wound. They can therefore, for example, indicate how a body was dragged along a path and in this scenario particles of soil, cement or vegetation may become embedded in the wound, thereby providing clues of where and when the event took place. Abrasions are a common feature of traffic accidents and can indicate how a body rolled or slid across the surface of a road. Cyclists and motorcyclists are vulnerable to this type of injury and refer to it as gravel rash.

'Crush abrasions' result from being hit with a blunt object with sufficient force to abrade the skin but not to overstretch and tear it. This can occur during a traffic accident from the impact of a car bumper or radiator grill or when a person is lashed with a plaited horsewhip. The object causing the crush abrasion can leave an accurate impression in the skin that allows future identification. A record of the wound can be made using vinyl polysiloxane and this is then covered with isocyanate resin to produce a positive replica. This can then be observed using scanning electron microscopy (SEM) to reveal much finer detail than is possible with the naked eye or conventional light microscopy (Rawson *et al.*, 2000). Hanging usually leaves a grooved wound in the neck referred to as a 'ligature furrow', although such marks might be faint or absent if a soft material, such as a ripped bed sheet, was used. In suicidal hangings, the ligature furrow seldom completely surrounds the neck but is angled with an obvious 'suspension peak' indicating how the body was hung (Figure 3.8). By contrast, when a person has been murdered by having a length of cord, wire or rope drawn tightly about their neck, the ligature furrow usually goes completely around the neck (Figure 3.9). There may be exceptions to this if a collar or other clothing has got in the way of the ligature or if the ligature was applied by pressure across the front of the neck with the assailant standing behind. However, there is a

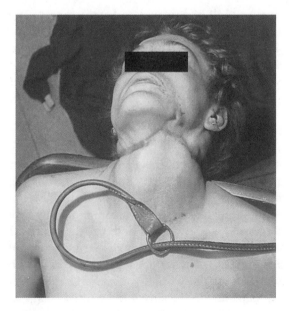

Figure 3.8 This man committed suicide by hanging himself with a dog lead. Note the ligature furrow around the neck, rising to form a suspension peak. (Reproduced from Shepherd, 2003, with permission from Arnold)

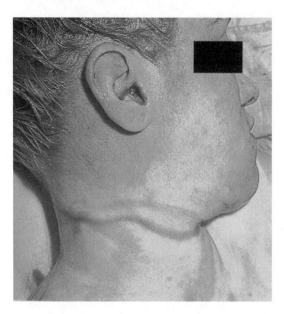

Figure 3.9 This man was strangled using a soft-silk ligature. Compare the wounds to those in Figure 3.8 and note how the ligature furrow encircles the neck without rising to a suspension point. (Reproduced from Shepherd, 2003, with permission from Arnold)

strong chance that there will be bruising and scratching from fingernails where the victim has attempted to remove the ligature and in neither case would there be evidence of a suspension peak. This method of killing is sometimes referred to as 'garotting' and a stick or similar device is used to tighten the ligature – this together with any knots in the ligature can cause localized bruising.

Lacerations

A blunt object hitting the surface of the skin with sufficient force to overstretch it and thereby cause it to split and tear causes lacerations (Figure 3.10). They are most commonly formed where there is only a thin layer of flesh above the bone, e.g. on the scalp and shins, and are accompanied by bruising or abrading of the surrounding tissue. The tissues underlying the skin surface vary in their strength and elasticity and consequently when tearing occurs not all of them are broken. This means that if the cut surface is examined tissue fibres will be found connecting the two sides of the wound – these are called bridging fibres. The amount of bleeding depends on the site of the wound (scalp wounds bleed profusely) and the size of the laceration. Where the force is applied tangentially, such as a glancing blow from an iron bar or in a traffic accident, large areas of skin may be torn (flayed) away. Lacerations seldom provide a good impression of the

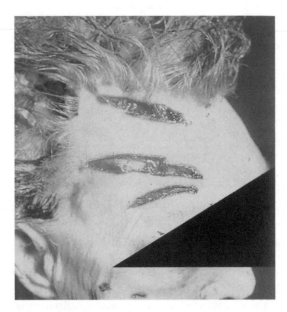

Figure 3.10 This man was savagely beaten with a heavy metal bar that has caused several deep lacerations. (Reproduced from Shepherd, 2003, with permission from Arnold)

object that caused them but, owing to the force involved, fragments may remain embedded in the wound. For example, a blow from a baseball bat or pickaxe handle might leave splinters of wood in the wound.

Incised wounds

Incised wounds (slashes or cuts) result from a sharp object cutting through the full thickness of the skin. They are distinguished from lacerations by the absence of bridging fibres and from stab wounds by being longer than they are deep (Figure 3.11). Typically, incised wounds are caused by knives or slashing weapons such as machetes although they can be produced by any sharp object such as a broken beer bottle, razor wire or even the edge of a piece of paper. In fights, slashing weapons tend to be wielded at arms length and blows are usually aimed at the face, neck and hands. If the cutting surface is extremely sharp, it will produce a wound with clean, well-defined edges. However, if the cutting surface is blunt then the wound will have abraded edges and the surrounding tissue exhibits bruising. In neither case would it be possible to deduce much about the nature of the weapon used beyond its sharpness and it is highly unlikely that it would leave trace evidence behind. An exception might be where the weapon was fragile, such as a broken glass or bottle, in which case fragments can break off in the wound. The start of the wound is usually the deepest

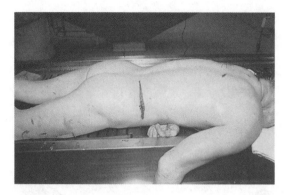

Figure 3.11 Typical incised wound that is longer than it is deep and inflicted by a slashing weapon. (Reproduced from Shepherd, 2003, with permission from Arnold)

and can therefore indicate the direction of the blow, e.g. whether the knife was pushed forward, pulled backwards or used in a sawing manner. The wounds can be replicated, after a fashion, in the laboratory using a ripe melon as the body and a variety of knives as weapons and comparing the 'wounds' they inflict. Unless a major blood vessel is damaged, or they are deep and extensive, incised wounds might bleed profusely but they are seldom life threatening. However, large slashing weapons, such as swords, are much more likely to inflict fatal cuts or even decapitate or amputate the limbs of the victim. Both incised and stab wounds to the back are always highly suspicious.

In the case of suicide, the victim often tests his or her personal resolve and the sharpness of the weapon on the fingers or makes one or more parallel tentative initial cuts to the wrist or neck before making the fatal incision. Suicides who cut their own throat usually begin the fatal cut high on the neck on the opposite side to the hand holding the knife. The weapon is then dragged downwards, passing through the common carotid artery, and may sever the hyoid bone and thyroid cartilage. The weapon will usually be found close by and may even remain in the victim's grip. Where the throat is cut during the course of a violent attack, the tentative incisions will be lacking although there may be more than one cut present. The cuts will usually be forceful and delivered across the front of the throat and may begin from either side depending on whether the assailant was right- or left-handed and the attack took place from in front or behind the victim. Typically, the larynx and all the major blood vessels will be severed. The diagnosis of suicide, especially by high-profile individuals, often creates controversy. For example, during the Hutton enquiry into the death of the UK government biological weapons expert Dr David Kelly, the official verdict was death by suicide as a consequence of blood loss following him slitting his left wrist. This view has been challenged by other doctors who consider that since only one artery – the ulnar artery – had been cut this would have been unlikely to lead to sufficient blood loss to be fatal (www.guardian.co.uk/letters/story/0,3604,1131833,00.html). When an artery is cut through completely, the thick vessel walls spontaneously contract and this reduces bleeding. This explains why people who lose a limb in an industrial accident or during a battle have been known to walk or crawl away, sometimes for long distances, and also, possibly, why Dr Kelly's body was surrounded by relatively little blood. It should also be noted that a person slashing their neck or wrists might die as a result of an air embolism rather than blood loss. In this case, air enters the cut ends of the veins and travels back to the right atrium of the heart where it causes a cardiac arrest. Similarly, when the throat is cut, death may result from asphyxiation when blood enters the lungs.

Healed incised wounds are often a feature of past medical operations, such as removal of the appendix or a caesarian, and this may prove useful as an identifying feature. The effectiveness of the healing process is affected by numerous factors, such as the nature of the wound, the part of the body, general health and age. Interestingly, wounds inflicted during early gestation will heal rapidly and perfectly. A patchwork of healed and partially healed incised wounds is a common feature of self-harm. These wounds can be anywhere on the body although they are usually in places that can be hidden by clothing and are always restricted to sites that the victim can reach. Wounds to the mouth, nipples, eyes and genitals are common in sex-related crimes when an assailant wishes to mutilate his victim, although persons suffering from extreme forms of mental illness will also harm themselves in this way.

Stab wounds

Stab wounds, also called puncture wounds, are those that are deeper than they are wide (Figure 3.12). Virtually any thin, rigid object – even a pencil or a toothpick – can produce a stab wound if it is wielded with sufficient force. For example, Lunetta et al. (2002) relate a case in which a young schizophrenic man stabbed himself so hard through an eye with a plastic ballpoint pen that it ended up lodged in his cerebellum. The man subsequently died of his wound but until he was autopsied it was assumed that he had been shot. Similarly, Grimaldi et al. (2005) describe a case in which a 72-year-old man committed suicide by stabbing himself in the chest using a pencil. Many weapons (e.g. knives, broken bottles) will produce both incised or stab wounds depending on how they are used. Obviously, the sharper the point, the easier it is for an object to cause a stab wound. Once the natural elasticity of the skin has been overcome, the rest of the implement will follow easily. This can be demonstrated, albeit crudely, by pressing objects of varying sharpness and diameter against a ripe melon. Placing an electronic pressure transducer underneath the melon will enable force comparisons to be made. Clothes will naturally offer some protection and buttons or objects carried in pockets can deflect blows and result in unusual wound tracks. They can also provide an indication of the victim's position when he was stabbed. Consequently, the body of a stabbing victim should always be

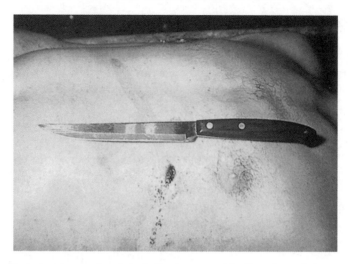

Figure 3.12 Typical stab wound and the knife that caused it. The edges of the wound have contracted and therefore make the wound appear narrower than the width of the knife. (Reproduced from Shepherd, 2003, with permission from Arnold)

examined initially with the clothes in place. This will enable the holes in the clothing to be compared to the underlying wounds. For example, if you raise your arms whilst wearing a coat or jacket, the material is also raised upwards. A blow to the upper chest at this time will pass through material that would cover the lower chest when the arms were held at rest. Consequently there will be an apparent discrepancy between the holes in the clothing and the stab wounds in the body. Similarly, the pattern of bloodstains on clothing is affected by the body's position. For example, blood seeping from a chest wound forms different trickle patterns depending on whether the victim is standing, sitting or lying on his back or sides. However, stab wounds do not always bleed profusely (at least not externally), especially if the implement is left in place.

Some criminals carry a sharpened screwdriver or chisel because, unlike knives, these are not immediately obvious as offensive weapons and can prove useful in breaking and entering premises. Transporting screwdrivers or similar tools in a haversack, especially whilst riding a motorbike, is tempting fate because in the event of a fall these could cause fatal stab wounds. Accidental stabbings are not unusual in kitchens and slaughterhouses and anywhere where knives are used routinely. They also happen when friends, usually youths, are fooling around with knives or other sharp objects. Such accidental stabbings usually involve a single puncture wound and if the victim was alone the cause of the wound will be found nearby (although he may stagger some distance before collapsing). Sometimes it is difficult to differentiate between an accidental stabbing and one in which an attack occurred, especially if there were no impartial witnesses. For example, in November 2000, a 10-year-old child, Damilola Taylor, was found dead in the stairwell of a south London housing estate. He had bled to death following a stab wound to his thigh that had severed an artery and cut through several veins. In a high-profile trial, four youths were charged with Damilola's murder. A medical trauma expert witness called by the defence stated that the wounds were more consistent with Damilola falling on a broken bottle than being intentionally stabbed. As mentioned above, severed arteries tend to constrict, but the defence claimed that by instinctively attempting to crawl home, Damilola may have opened the wound. By contrast, other medical witnesses suggested that Damilola had been stabbed with a broken bottle that was then twisted in the wound either deliberately, or as a consequence of the victim's movements. Ultimately, owing to a lack of forensic evidence linking the youths to the scene, they were acquitted but the case continued to cause controversy. In January 2006, the case returned to the courts when three different youths were charged with Damilola's murder. Although they were initially suspected of involvement in Damilola's death, they were never charged. According to the prosecution, vital evidence was missed during the initial forensic investigation. This evidence included a drop of Damilola's blood on a white Reebock trainer belonging to one of the defendants, and a second drop on the cuff of a sweatshirt belonging to another defendant. In addition, fibres from Damilola's clothes were found on all three defendants. This case indicates the value of storing and re-evaluating forensic evidence.

A few years ago, a notorious case of accidental stabbing took place in Leiden, The Netherlands, that demonstrates the importance of an experimental approach to wound analysis and the need to maintain an open mind (Rompen et al., 2000; Bal, 2005). In May 1991 a 21-year-old student returned home to find his mother dead. An autopsy demonstrated that a plastic Bic® biro had penetrated her right eye, entered her brain and lacerated the brainstem, causing her instant death. An unusual feature of the wound was that it had perforated her eyeball; in most similar eye injuries the eyeball is pushed aside by the penetrating object. However, two medical experts stated that the most likely scenario was that the woman had stumbled whilst holding the pen and the fall had forced it through her eye and into her head. The police did not believe this but in the absence of further incriminating evidence the case was dropped.

After about 4 years, a witness came forward claiming that the student may have killed his mother by firing the pen from a crossbow, whilst his psychologist (disregarding the supposed requirement to provide her clients with confidentiality) stated that he had confessed to the killing. The student was therefore arrested and at the trial he was found guilty and sentenced to 12 years in prison. An appeal was launched and, as part of this, experiments were undertaken to determine whether it really was possible to shoot a plastic ballpoint pen from a crossbow with sufficient force through the eye and into the skull. Initial experiments used the heads of dead pigs and a crossbow with a 10.9 kg tractive power. Unfortunately, these studies suffered from practical problems but it was demonstrated that, even with the crossbow held in contact with the eye, the pen did not penetrate the eyelid or the orbita. The researchers therefore obtained ethical approval to use two human corpses that were destined for dissection training. The crossbows were held in contact with the eyes, which were open, so that the pen would strike the eye with maximum force. However, no matter what angle the crossbow was fired from, and even if a bow with a tractive power of 40 kg was used, the ballpoint pens never penetrated as far into the head as the one retrieved from the dead woman. Furthermore, the pens fired from the cross-bows, with one exception, acquired an indentation mark from the drawstring and were damaged, a common feature being the central ink tube plus cone head telescoping from the shaft. By contrast, the pen retrieved from the woman's skull was undamaged and had no indentation mark. These results, together with other evidence, led to the student's release and subsequent acquittal.

In addition to accidents, single stab wounds are also a feature of suicides and homicides in which the victim was unable to resist owing to being asleep or incapacitated in some way, such as through drink, drugs, illness or old age. These single homicidal wounds are usually delivered with care and precision. By contrast, if the homicide victim is conscious and able bodied, the first blow is seldom delivered so effectively. Even if the first blow is potentially fatal, it is unlikely to immediately incapacitate the victim and he or she attempts to either fight back or flee. In such cases there will probably be defence wounds to the arms and hands as the victim attempts to ward off further blows. Sometimes the victim may grasp the blade, resulting in incised wounds to the palms and fingers. It is not possible to state how long a victim would remain alive after receiving a fatal wound. There is a great deal of variation between individuals and sometimes a victim may continue to resist or run for several minutes after being subjected to a series of potentially fatal blows. Multiple stab wounds are also a common feature of sexually motivated attacks and may be accompanied by mutilation. Sometimes there are large numbers of wounds as the assailant becomes frenzied and stabs his victim wildly. Altogether, sharp instruments are responsible for the majority of homicides in England and Wales. In recent years, the figures have varied from 27 to 33 per cent of all homicides, and the pro-portion is higher in men (28–35 per cent) than women (23 to 31 per cent) (www.crimereduction.gov.uk/statistics38.htm).

In suicidal stabbings, the victim often removes the clothing around the chosen wound site. This region is known as the 'elective' site or area and is commonly in the area of the precordium (i.e. that overlying the heart and stomach) or the abdomen – and the wound is directed upwards aiming for the heart. The knife may be partially withdrawn and reinserted several times and there might be more than one stab wound. However, there will not be any defence wounds and there may be evidence of tentative incised wounds either to the fingers or in the elective area where the victim tested the sharpness of the weapon. Sometimes the suicide victim may attempt to stab himself through the neck (rather than slitting the throat) and interpretation of such wounds is very difficult because this is also a very efficient way of murdering someone. The majority of suicides do not leave notes explaining their actions and even if a note is found the pos-sibility that it was written under duress should be considered. A stabbing victim may also be forced to stab himself during a struggle (e.g. the hand grasping the knife is turned against him) but in such circumstances there will probably be extensive bruising elsewhere on the body.

The size and shape of a stab wound are not always an accurate reflection of the implement that caused it. For example, the victim may attempt to twist away when the blade is inserted, thereby widening the wound and giving it a triangular profile. Similarly, the assailant may rock the blade to increase penetration. In skilled hands this technique can have dramatic consequences: slaughterhouse workers use it to eviscerate an animal and can split the ribcage of a large pig from base to top in a single movement. The handiwork of a person trained in the use of knives and butchery or medicine will therefore be distinct from the crude hacking of someone overcome with anger and emotion. Another complicating feature in the interpretation of knife wounds can be the contraction of elastic fibres in the skin surrounding the wound. This can result in the wound being deformed and make it look as though more than one weapon was used in the assault. Similarly, it is possible for a single thrust to cause more than one wound if it passes out of flesh and then back in again. This can happen if the knife is being thrust at a shallow angle in relation to the body or passes through thin regions of flesh (e.g. the blade perforates the hand and passes into the chest). The depth of a wound may exceed the length of the blade that caused it. This is because some areas of the body, such as the abdomen, can be compressed and therefore once the blade has been inserted up to its hilt one can continue pressing inwards. This is particularly the case if the victim is held tightly or forced against a solid object such as the floor or a wall. It can mean that an impression of the hilt is left at the wound site in the form of a bruise or a transfer bloodstain. Relatively blunt stabbing weapons, such as an unsharpened screwdriver, cause the skin to spit around the point of insertion and there may be associated bruising and laceration of the surrounding skin. Single-bladed knives, such as carving knives, often produce a wound in which one end has a sharp V-shaped profile and the other has a blunter end that has split and produced a triangle or spike: this is called 'fish-tailing'. Knives in which one edge is serrated, such as those used by fishermen, will produce wounds in which one edge is torn and lacerated. Stab wounds to the soft tissues may be difficult to identify once a body has entered the late stages of decay and once the body is skeletonized it becomes impossible. However, if during the course of the attack any of the bones were chipped – as is highly likely if the wound was to the chest – then this can raise suspicions about the cause of death (Bonte, 1975). Observing scratches, chips and other tool marks on bone is not always easy with the naked eye. Consequently, where there is doubt, SEM can be used to reveal fine detail (Alunni-Perret *et al.*, 2005).

Bone damage

The skeleton remains long after the soft tissues have decayed and can provide valuable information on the cause of death. (Quatrehomme and Iscan, 1997a)

Similarly, evidence of bones that were broken some time in the past but had since healed can be a useful identifying feature. It takes considerable force to fracture or break the bones of an adult provided that they are not afflicted by a medical condition such as osteoporosis. Consequently, the presence of broken bones in a dead body always arouses suspicions. For example, fracturing of the cervical vertebrae is unusual in suicidal hangings unless there was a long drop and the noose was prepared properly: most suicidal hangings result in death by asphyxia. Consequently, an otherwise healthy adult male who had died from a broken neck may represent a case of homicide rather than suicide if he was found suspended from a length of electric cable attached to a low branch. However, if the noose is composed of material that is both strong and relatively inelastic and the victim falls a suitable distance before it tightens, then it is possible for him or her to be decapitated. The distance is very dependent upon the weight of the victim: the heavier the victim, the less distance he needs to fall. It was part of the skill of a professional hangman to judge the drop required to avoid slowly throttling him at one extreme and beheading him at the other. Tracqui *et al.* (1998) relate a case in which a young man jumped from the bridge above a canal with a nylon rope around his neck. He fell for 3.7–5.3 metres before the rope tightened and when it did he was instantly decapitated, his head sinking beneath the bridge and his body drifting over 200 metres downstream.

It should be remembered that a broken bone might result from either direct force or forces transmitted from elsewhere in the body. The latter is often the case in the thorax where a single blow may result in multiple fractures within the ribcage as the forces radiate outward and create alternating points of tension in which one fracture can beget another (Love and Symes, 2004).

The hyoid bone forms part of the axial skeleton and two characteristics make it unusual (for a bone): it is a single U-shaped bone that does not have a partner, and it does not articulate with any other bone. It is found in the anterior region of the neck between the mandibles and the larynx and its function is to act as a sling to support the tongue and for some of the neck and pharynx muscles (Figure 3.13). Damage to the hyoid bone, especially one or both of the horns of the 'U', is a characteristic sign of manual strangulation. It is far less likely to occur in hanging owing to the positioning of the rope and can therefore help to distinguish between suicide by hanging and manual strangulation. It is, however, important to remember that absence of damage to the hyoid bone does not necessarily mean that strangulation did not take place. A great deal depends on how and where pressure was applied to the neck. Similarly, in adults, manual strangulation is likely to result in fracturing of the thyroid cartilage (Adam's apple) but this may not occur in young children because their cartilage is more pliable. Many more women are killed by strangulation than men because there needs to be a considerable difference in the relative strength of the assailant and victim for it to be successful. In England and Wales, 15–25 per cent of all female homicides in recent years have resulted from strangulation (including asphixiation) but for men the figures vary from only 3 to 7 per cent (www.crimereduction.gov.uk).

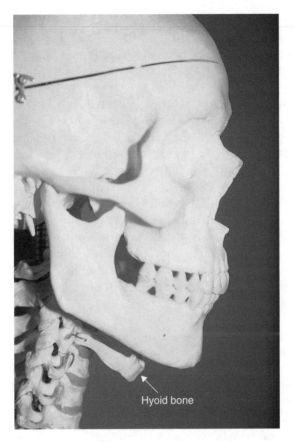

Hyoid bone

Figure 3.13 The position of the hyoid bone can be clearly seen on this plastic model of a human skeleton

Gunshot wounds

Gun crime has increased enormously in England and Wales in recent years: in the year to December 2004 there was a provisional total of 11 082 firearm offences. However, a significant proportion of these involved the use of imitation weapons and relatively few were homicides. For the year 2002–2003, the most recent for which figures are available, only about 8 per cent of all homicides (80 out of 1007) involved the use of firearms – and this includes the use of the firearm as a blunt instrument to bludgeon the victim to death (www.crimereduction.gov.uk).

There are many different sorts of firearm and ammunition and each results in slightly different wound characteristics. In addition, there are many complicating factors that influence the nature of the wound. Consequently, only a basic coverage is provided here and those requiring more detail are advised to consult DiMaio (1999). Gunshot wounds are penetrating injuries that involve an entrance hole and possibly also an exit hole. The latter situation is most likely to occur if the bullet was fired from a high-velocity gun or from close range. Sometimes the bullet will travel through the body in a straight line but if it hits a bone or it starts to tumble badly then its trajectory can be changed. For example, a bullet that enters the chest may leave via the neck or come to rest in one of the legs. If an exit hole is formed, it is typically larger than the entrance hole and bleeds more profusely. This is because when a bullet enters the body the skin initially deforms and then, once the bullet has pierced and passed through, it retracts back. Consequently, the entry wound may appear to be smaller than the diameter of the bullet. As the bullet passes through the body, even if it continues in a straight line, it will usually start to tumble and it therefore carves out an exit hole that is at least as large as its diameter and it might force out fragments of tissue and bone. If there is some impediment to the bullet leaving the body, e.g. the body is pressed against a wall or the floor, or the victim wears a stiff trouser belt, then a 'shored' wound is formed in which the edges of the wound are abraded against the overlying object.

The range from which a victim was shot can often be judged from the nature of the wounds. If the gun is pressed firmly against the flesh it forms a 'hard contact wound'. The edges of the wound are seared and blackened by the heat of the explosive charge coupled with soot and propellant being baked into the skin (Figure 3.14). With no way of escaping, the hot gases from the discharge are forced into the body, causing the surrounding tissues to balloon and the skin to split. The latter is most likely to occur in shots to the skull and wherever else there is bone directly underlying the skin. The high temperature of the gases burns the surrounding tissues, so blackening is found along the wound track. The wound will also contain particles of soot and unburnt powder and, in the case of smooth-bore shotguns, fragments of wadding. The presence of carbon

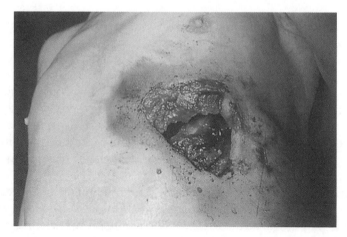

Figure 3.14 A shotgun was discharged close to this man's chest. Note the massive disruption caused by the entry of gas and shot into the skin. There is a single gaping wound, the edges of which are blackened with soot. (Reproduced from Shepherd, 2003, with permission from Arnold)

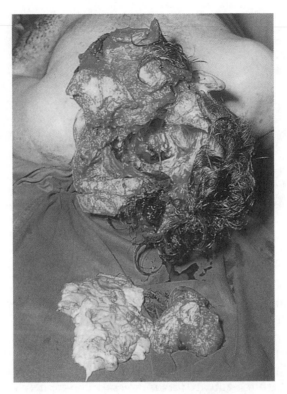

Figure 3.15 When a shotgun is fired into the head, the forceful entry of hot gas and shot into the cranium can cause the skull to literally explode. In this case the suicide victim placed the barrel into his mouth and then pulled the trigger. (Reproduced from Shepherd, 2003, with permission from Arnold)

monoxide (formed by incomplete combustion of the explosive charge) can cause the surrounding tissues to turn pink. The carbon monoxide binds to the haem part of the blood pigment haemoglobin to form carboxyhaemoglobin and this is responsible for the cherry pink coloration. The consequences of gases being forced into the body are particularly dramatic with smooth-bore shotguns and if these are discharged into the head it can result in the whole skull splintering and exploding (Figure 3.15). This commonly occurs in suicides where the muzzle of the gun is placed in the mouth or underneath the chin. Where such appalling injuries are caused, it is essential to examine the whole body and the room in which it was found because it is possible that the person was murdered and then a fake suicide staged to mask the evidence. For example, was it possible for the victim to reach the trigger of the gun when holding the muzzle to his head? If his arms were not long enough, the victim will sometimes use his toes or a prop to pull the trigger but if he is fully shod and there is no prop then suspicions should be raised.

If the gun is held at a slight distance from the body, the gases are able to escape, so the damage caused by ballooning is not seen. The circumference of the black-

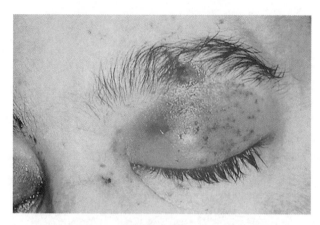

Figure 3.16 Powder tattooing around the entry wound indicate that the pistol was fired at close range. The bruising results from the skull being fractured by the bullet. (Reproduced from Shepherd, 2003, with permission from Arnold)

ening (smoke soiling) around the entry wound initially increases and then becomes more diffuse as the distance increases. Unlike hard contact wounds, it is possible to wipe away this blackening. The pattern formed by smoke soiling can indicate whether the gun was held at right angles or obliquely to the wound site. Smoke soiling usually ceases at a distance of about 1 metre. Among the gases discharged from a gun are tiny fragments of extremely hot propellant and when they come into contact with the skin they cause a characteristic pattern of burns called 'powder tattooing'. The diameter of the powder tattooing indicates the distance between the victim and the muzzle of the gun but, like smoke soiling, it is not normally found beyond one metre (Figure 3.16). The colour of the powder tattooing can indicate whether the victim was alive or dead at the time of the shooting. If the victim was alive, the marks tend to be reddish brown to orange red, whereas if he were dead the marks would be grey or yellowish.

If the victim was shot through clothing it can affect the appearance of the wound. For example, soot and fragments of propellant may not reach the skin and consequently there may not be any smoke soiling or powder tattooing. However, marks may be found on the clothing and fragments of clothing material might be found along the wound track. As with stab wounds, aligning the entry holes in the clothing with those in the body can indicate the position of the victim (i.e. sitting, standing, arms raised, etc.) at the time he was shot. Obviously, a great deal depends on the nature of the clothing, the distance and the weapon that was used. In addition, the use of silencers may produce unusual entry wound characteristics.

Shotguns usually fire lead or steel pellets (shot), the size of which is denoted by numbers (e.g. 5, 6, 7.5, 8), with the larger shot being used for bigger game although they are also used to fire single heavy projectiles called slugs. Shotgun pellets spread out as they leave the muzzle of the gun although if it is discharged close to the body they will cause just a single large entry wound. At a distance of 20 cm or more, the pellets cause separate entry wounds and the spread can indicate how far away the muzzle was from the victim. However, there is a great deal of variation between guns and the type of shot used also affects the pattern. It is therefore essential to conduct tests on any gun that might have been used in a crime before firm conclusions are reached. Shotgun wounds resulting from a gun fired from less than about 3 metres away will commonly include fragments of wadding. This can be very useful in determining the type of shot and the gauge of the gun that was used. With increasing distance it is less likely that any pellets will pass through the body and above 20–50 metres the body might be peppered with shot but the wounds are unlikely to be fatal.

'Rifled weapons' is the term used to include a diverse selection of firearms that are sub-divided into three categories: long-barrelled rifles, short-barrelled revolvers and automatic pistols. Rifles tend to have greater accuracy and discharge their bullets with greater velocity than revolvers and pistols but there is a great deal of variation between weapons and a wide variety of ammunition. The wounds they cause depend on the amount of energy the bullets impart to the body and this in turn depends on the type of weapon and hence the velocity and calibre of the bullets used, the distance of the victim from the muzzle of the gun and the stability of the bullets' flight. Obviously, a bullet travelling at high velocity imparts more energy, and therefore causes more damage, than the same sized bullet travelling at low velocity. Similarly, a heavy-calibre bullet will cause more damage than a small-calibre bullet travelling at the same speed. Both low- and high-velocity firearms cause penetrating injuries resulting in direct damage to the tissues as the bullet traverses the body. Because of their greater kinetic energy, bullets from high-velocity firearms are more likely to pass through the victim and for this reason the armed police patrolling English crowded city centres are often issued with low-velocity firearms to reduce the risk of hitting innocent bystanders. In the case of low-velocity weapons, the bullet forms a permanent cavity that is an accurate reflection of the amount of damage caused. By contrast, a bullet from a high-velocity firearm possesses far more kinetic energy and when it hits a body it causes the formation of a temporary cavity that is much wider than the diameter of the bullet. This temporary cavity lasts only a few milliseconds and travels through the body with the bullet, collapsing behind it in a series of pulses. This results in a wide area of bleeding around a much smaller permanent cavity that may be little wider than the diameter of the bullet. The consequences of the pressure waves vary between tissues: the denser the tissue, the easier it is for the bullet to transfer its kinetic energy and therefore the greater is its potential to suffer damage. However, the risk of damage is reduced if the tissue is able to dissipate the absorbed energy through expansion and contraction. The tissues of the lungs, having relatively low density and also being highly elastic, are therefore the least likely to suffer harm in this way. Similarly, muscle tissues, although they are dense, are highly elastic and are therefore protected to some extent. By contrast, dense, inelastic tissues, such as the brain, spleen and liver, may burst, because they are unable to expand and contract as the pressure waves pass through. Similarly, pressure waves can cause bones to shatter whilst the large intestine may rupture as the gases within it expand. If they hit bone, bullets from both low- and high-velocity firearms are likely to cause it to fragment. This results in secondary projectiles within the body that cause their own wound tracks. It is also likely to alter the trajectory of the bullet itself. Some workers have attempted to use the direction of the bevelling produced in bones as the bullet exits from them as a means of estimating the direction of fire but this is not particularly reliable (Quatrehomme and Iscan, 1997b).

If a person is shot with a rifled firearm from a distance of over 1 metre but still at relatively close range, the entry wound will probably be more or less circular and surrounded by an 'abrasion collar'. This abrasion collar results from a combination of dirt, heat and friction from the bullet as it initially pushes the skin inwards before penetration occurs. The precise shape of the entry hole will be affected by the angle with which the bullet struck the skin, and this will be affected by the position of the muzzle in relation to the body. As the distance between the muzzle of the gun and the victim increases, there is an increased chance of the bullet starting to yaw or tumble before it reaches its target. When such a bullet arrives it will tend to cause a P- or D-shaped entry wound and lacerations of the surrounding tissues so that it might be mistaken for an exit wound. However, unless they are shored (see above), exit wounds are unlikely to have an abrasion collar and will be everted rather than inverted. If the bullet becomes deformed or fragments on impact, huge gaping wounds can result. A similar situation might occur if the bullet had ricocheted or already passed through another body. Dum dum bullets are ones that are doctored so as to make them prone to expanding and disintegrating on impact, thereby causing huge injuries (Sykes *et al.*, 1988). They get their name because they were developed at a British army cantonment based in the region of Dum Dum on the outskirts of Calcutta (modern-day Kolkata) in the nineteenth century (Spiers, 1975). Such bullets were banned by the first International Peace Conference at the Hague in 1899 and also at the 1933 Geneva Convention. Although the ban is still in force, the reason why dum dum bullets are not used more widely probably has more to do with their tendency to jam in self-loading firearms and the development of high-velocity rifles that can cause equally devastating injuries. It has been alleged that dum dum bullets were used in the Bloody Sunday deaths, although this has not been proved conclusively because some of the bullets could already have been tumbling as a result of ricochets before they hit their victims, hence causing more serious injuries than would have been expected (www.birw.org/bsireports/51_70/report64.html). (On 30 January 1972, British soldiers in Northern Ireland shot dead 13 men and injured 14 others who were taking part in an illegal demonstration in the town of Londonderry. A further man died later of his wounds. The event has been known ever since as 'Bloody Sunday'.) Similar controversies exist in Israel where Palestinians have accused the Israelis of using dum dum bullets. The Israelis claim that the devastating wounds seen on some victims are a consequence of the bullets being fired from sniper rifles using 7.62 mm ammunition rather than the 5.56 mm ammunition used in the standard M-16 assault rifle. However, the bullets used in the M-16 have a reputation for tumbling and fragmenting when entering the body. It is therefore not necessary to resort to using dum dum bullets in order to produce equally horrific injuries.

Differentiating between suicidal, accidental and homicidal gunshot wounds is not always easy. For example, in the case of suicide and accident, the weapon may be flung far from the victim's grip. Furthermore, if there was some time before the authorities discovered the victim, the weapon may be missing because it was stolen or hidden by someone else. This might happen because a relative considered suicide to be a sin and therefore should be 'covered up', or a verdict of suicide might negate a life insurance claim or the firearm may have 'resale value'. These possibilities should be considered on the basis of the situation in which the body was found. The presence of powder tattooing can prove useful in determining the distance from which the person was shot and hence whether suicide was physically possible. The number of gunshot wounds is not always a good indication because suicidal individuals have been known to shoot them-selves more than once. However, the location of the wounds is more helpful (see Table 3.3). Suicides tend to shoot themselves in the head, usually in the right temple if it is with a handgun – the choice of temple tends to be determined by whether or not a person is right- or left-handed but it is not an infallible pre-dictor. Chest wounds may also result from suicide attempts. Gunshot wounds resulting from an accident can appear virtually anywhere, especially if the bullet ricochets before entering the body.

When a person handles a gun he will leave behind his fingerprints; if he fires the gun, then his hands and clothes will be contaminated with gunshot residues. These residues contain metals such as lead, antimony and barium that can be detected by flameless atomic absorption spectrophotometry (FAAS) and scan-ning electron microscopy energy-dispersive x-ray spectrometry (SEM-EDX). Firing a gun will usually result in residues being found on the backs of the hands and the palms, whereas merely handling a gun that was recently fired will trans-fer residues solely to the palms. The latter situation may happen if the weapon is passed to an accomplice or if the victim holds the barrel of the weapon in an attempt to deflect it (or, in the case of a suicide, to steady it). The residues are rapidly lost, especially if the hands are washed. Similarly, some firearms (e.g. certain rifles) deposit low amounts of residues, which may lead to false nega-tives, whilst others, such as revolvers, produce large amounts that may even contaminate bystanders and lead to false positives. Further potential problems with gunshot residue analysis arise from a variety of angles (Mejia, 2005). For example, it is claimed that a person entering a room shortly after a shot was fired in it may be contaminated with more residues than the person who pulled the trigger and promptly fled (Fojtasek and Kmjec, 2005) and that particles with a similar composition to gunshot residues can be formed by fireworks and within brake linings. In the latter situation, it is suggested that a person servic-

Table 3.3 Distinguishing features of wounds associated with suicide and homicide

Wound type	Suicide	Homicide
Bruise	Not commonly seen in suicide attempts	Common. Distribution provides evidence of assault and restraint
Abrasion	Typically associated with hanging: a suspension point is usually obvious and there is no evidence of a preceding struggle	Common. Caused during a struggle or when a body is dragged. Garrotting with a ligature: abrasion encircles neck and there is no suspension point
Laceration	Not commonly seen in suicide attempts[a]	Common. Caused by being hit with heavy blunt object
Incised	Slashing of wrist or throat. Evidence of tentative initial cuts at elective site. Incised wounds not found elsewhere unless past history of 'cutting'	Common. May occur anywhere on body but often aimed at face and neck. Defence wounds may be found on hands and arms. Distribution indicates how assault progressed
Stab	Typically aimed at heart. Clothing usually removed around wound site. Tentative initial cuts to test sharpness may be present. Multiple stab wounds possible	Common. May occur anywhere on body. Stab wound delivered through clothing. Number may vary from one wound to many in a frenzied attack. Defence wounds may be found on hands and arms. Distribution indicates how assault progressed
Broken bones	Not commonly seen in suicide attempts[a]	May occur during violent physical assault or shooting
Gunshot	Typically contact or very close range and aimed at head or chest. May be more than one wound	May be contact or from a far distance. Wound may occur anywhere on body. Wound analysis indicates position of assailant and victim
Burns and scalds	Self-immolation is not a common form of suicide. However, localized burns and scalds may be found on a suicide victim from previous self-harm. Chemical burns to mouth and oesophagus are caused by suicidal ingestion of bleach or caustic soda	Localized burns and scalds may result from torture before victim was killed. Burning and scalding not a common cause of homicide. Burning is a common means of disposal of a dead body. Presence of soot in lower respiratory tract and >10 per cent carboxyhaemoglobin in blood indicates that victim was alive when fire commenced

[a]Suicide resulting from jumping from a tall building or in front of traffic or a train results in severe abrasions, lacerations, broken bones and even amputations.

ing the brakes of a car may acquire residues on his or her hands and clothing that could be mistaken for gunshot residues (Cardinetti *et al.*, 2004).

Taking DNA samples from the bloodstains found on knives, hammers, etc. has been common practice for many years but a new field of research is being developed called bullet cytology (Knudsen, 1993) in which tissue samples and DNA are recovered from bullets. Bullets from high-velocity guns often pass straight through a victim, so this technique could prove useful in linking the two together. However, there have been few publications on the topic.

Bite marks

Humans bite one another with remarkable frequency and they are a common feature of physical and sexual assaults. Both the victim and the assailant are likely to bite one another although there are differences between the locations in which bites are inflicted. In a survey of biting injuries reported in court cases in the USA (Pretty and Sweet, 2000) 17 per cent of bite marks were found on men, of whom 52 per cent were the accused, whilst 83 per cent were found on females, none of whom were the accused. Male victims of crime tended to be bitten on the arms (27 per cent) and back (9 per cent) whilst the assailant tended to be bitten on the hands (18 per cent) and arms (36 per cent) (all percentages are a proportion of all male bites). Female victims tended to be bitten mainly on the breasts (40 per cent), arms (13 per cent), legs (13 per cent), face (7 per cent), neck (7 per cent) and genitalia (7 per cent). In particularly violent attacks, parts of the ear, nose and other facial features may be bitten off entirely. Assuming that the dental characteristics are rendered in sufficient detail at the bite site to enable identification, bite marks can link victim and suspect together. Unfortunately, this is seldom true and Pretty and Turnbull (2001) relate a case in which an assault had taken place in the course of which the victim had been bitten. Although the dental characteristics of the two assailants were different, they were also sufficiently similar for two forensic dentists to arrive at opposite conclusions about which of them had inflicted a bite mark. Contradictory views like this are not unusual and there is a need to improve the ease and accuracy of bite mark analysis. Obviously, DNA analysis of saliva left at the bite site can provide an accurate identification of the assailant but it is a natural reaction of the victim (if alive) to immediately wash the wound and even if this is not the case the DNA may degrade before it can be analysed.

One of the problems of bite mark analysis is that bites inflicted in anger are painful and unwonted and the victim therefore attempts to pull away. This results in the teeth being dragged through and over the skin so the marks are not clear. Furthermore, the bites are inflicted on three-dimensional, curved surfaces of soft skin but the comparisons of bite site and dental characteristics are usually done from two-dimensional photographs. A further complication is that bites inflicted whilst the victim is alive cause bleeding, bruising and swelling and this can distort the bite site. However, these problems could be at least partially overcome through the use of computer-based image capture systems that enable three-dimensional records to be made (Thali *et al.*, 2003).

Not all bites are the consequence of assault. Love bites on the neck are very common, especially among teenagers. These, and other amorous bites, are usually delivered slowly and the recipient does not draw away so aggressively. Consequently, there is not the evidence of dragging. However, during these bites, the skin is often sucked in and the tongue thrust forcefully against it, thereby causing the formation of tiny red spots called petechiae in the centre of the bite that result from blood leaking from damaged blood vessels. Furthermore, there may not be any evidence of tooth marks. Self-inflicted bites are commonly found on persons who become addicted to self-harm and may show similar characteristics. These bites will, however, not occur on areas of the body that are difficult to reach and there will probably be wounds of varying ages. Nevertheless, it is feasible for someone who bites himself in this way also to be attacked and bitten by someone else and wounds of varying ages are often a feature of long-standing abuse.

Burns and scalds

Medically, a burn is defined as a wound that results from dry heat, such as touching a hot plate, whilst a scald results from wet heat, such as exposure to steam or molten metal. However, in colloquial speech, a burn is often used to describe the wounds that result from all these situations.

There are several classification systems for categorizing the severity of burns, although the most commonly used is to divide them into three 'degree categories'. 'First degree burns' are those in which there is reddening of the skin and a fluid-filled blister may form. These superficial burns usually heal without leaving any scarring within 5–10 days. In 'second degree burns' the epidermis of the skin is removed down to the dermis, whereas in 'third degree burns' tissue lying beneath the dermis is also damaged, the underlying organs may be exposed and there may be blackening resulting from carbonization. Scalds can be categorized on a similar basis although carbonization would only result from exposure to molten metals or similar high-temperature fluids. In terms of survival, the extent of the burns or scalds is of as much importance as their category. If over 50 per cent of the body surface is affected the chances of survival are poor, although it is difficult to generalize as there is a great deal of difference between individuals, and factors such as age, health and the speed with which medical assistance is administered all play a part in recovery.

The majority of people who die in house fires succumb to the inhalation of smoke and toxic fumes rather than being burnt to death. The presence of soot in the mouth and nasal passages is not a reliable indication that a person was alive at the time a fire was started because it could have entered passively. By contrast, the presence of soot in the lungs and air passages below the vocal cords is a very good indication, as is a saturation level of carboxyhaemoglobin in the blood in excess of 10 per cent. Carboxyhaemoglobin results from the inhalation of carbon monoxide produced by incomplete combustion of flammable materials – the blood and tissues appearing cherry pink intimate the presence of large amounts of carboxyhaemoglobin in the body. However, the level of carboxyhaemoglobin needs to be confirmed by chemical analysis because low levels (less than 10 per cent saturation) might occur naturally from living near to a pollution source or from being a heavy smoker. It is also important to remember that being alive is not the same as being conscious and the victim may have died before the flames reached him or her. Similarly, the absence of either of these indicators does not mean that the victim must have been already dead: a great deal depends on the individual circumstances and the amount of smoke and carbon monoxide produced. It is very difficult to distinguish between burns that were caused before, at the time of or after death. In the case of flash fires or explosions, in which death may be virtually instantaneous, there might be extensive surface burning but little evidence of soot in the lungs or carboxyhaemoglobin in the blood.

A body exposed to high temperatures automatically adopts a characteristic 'pugilistic posture' in which the limbs become flexed. This occurs regardless of whether the person was alive or dead at the time of exposure. As the skin and underlying tissues dry out, they contract and splits occur that can be mistaken for wounds. The drying out of the muscles induces them to contract and because the flexor muscles are larger and more powerful than the extensor muscles they pull the body into the 'pugilistic posture'. The bones can fracture and break and a 'heat haematoma' may form in the skull, leading to suspicions that the victim may have been assaulted. A haematoma is a blood clot that forms outside of a blood vessel and this is often a feature of a skull fracture or a serious blow to the head.

If a body is burnt, it will inevitably destroy a great deal of forensic evidence. Consequently, it is common for a murderer to either remove the body and burn it elsewhere or burn it *in situ*. An examination of the scene is therefore essential in determining the likelihood of a body being the victim of a suicide, a homicide or an accident. It is rare for persons to commit suicide by burning in the UK, although it does happen occasionally. Usually, the victim would have covered him or herself in a flammable liquid, residues of which would be detectable, a means of setting fire to themselves would be found nearby and there would be no evidence of restraint or injuries caused before death. In the case of accidental death, there would be evidence of an innocent cause, such as a chip-pan fire or smoking in bed, and that the person had attempted to escape unless incapacitated through age, a medical condition or intoxication. This would be indicated by the position of the body – usually the victim attempts to reach a door or window or, if cut off, to hide in a corner or cupboard. In a homicide, the fire would be started deliberately and the victim may, if not already dead, be restrained in some way, such as being tied to a chair or confined to the boot of a car. The distribution of burns on the body and clothing can also provide an indication of the person's position. For example, extensive burns to the top of the head and face would be suspicious in a reported kitchen accident involving clothing being set alight. In this scenario, the possibility that a flammable liquid may have been poured over the victim and then set alight should also be considered.

Burns and scalds are a common feature of self-harm, the abuse of children and old people and deliberate torture (Greenbaum et al., 2004). Both self-harm and the abuse of others often occur over a long period of time, so there may be wounds of varying stages of healing. The distribution of the wounds will indicate whether the victim could have caused them him or herself. Scalds often form trickle-shaped wounds indicating the flow of the fluid and hence the orientation of the victim at the time. Torture with hot implements usually takes place over a restricted period and is often accompanied by physical beating. If the implement is dry, such as an iron, its shape will be reproduced in the burn. The use of electric shock torture is difficult to demonstrate because, although it is extremely painful, provided that the electrodes are not applied for too long there is little associated pathology.

Postmortem injuries

Postmortem injuries can result from both malicious and accidental acts and if there were no witnesses to the person dying their interpretation can present difficulties. For example, an elderly person suffering a fatal heart attack could collapse and suffer serious injuries when hitting the furniture or the floor. On discovery of the body it would be necessary to exclude the possibilities that the victim either first received a blow to the head or was thrown to the ground by an intruder and the shock had induced the heart attack. This scenario might be resolved by a combination of bloodstain pattern analysis (Chapter 2), fingerprinting and wound analysis.

Injuries caused days or months before death are usually recognizable from signs of healing but it is difficult to generalize about postmortem injuries (Byard et al., 2002). Postmortem bruising may occur when a dead body is being recovered or handled in the morgue, although these bruises tend to be small. Postmortem bruising is most likely to occur on the lower body surfaces because this is where blood pools. Decaying soft tissues are very delicate and easily damaged so it is normal to undertake a preliminary examination before the body is moved to the morgue. Clumsy handling can result in limbs becoming detached and bones broken. Vertebrates can cause extensive postmortem injuries up to and including consuming the whole corpse (Chapter 8). Invertebrates can also cause damage that may result in confusion at the time of autopsy. Feeding by dermestid beetles and their larvae produces irregular holes and tears in mummified skin that can be mistaken for wounds caused before death. These holes tend to have jagged edges, whilst those produced by maggots tend to have smooth edges. Both histerid beetles and burying beetles (Silphidae) will cause circular-shaped holes in a body. Bite marks inflicted by the cockroach *Dictyoptera blattaria* have been mistaken for evidence of non-accidental injury or even strangulation in cases of sudden infant death (Denic *et al.*, 1997). Ants have been reported to cause postmortem damage that resembles premortem strangulation or burns caused by strong acids (Byrd and Castner, 2001; Campobasso *et al.*, 2004), although because the marks are inflicted after death they are not usually associated with bleeding.

Quick quiz

1. Distinguish between plastic, visible and latent fingerprints.

2. Explain why you would not use silver nitrate to detect fingerprints from windows of a house overlooking the sea.

3. Briefly describe how tattoos can provide clues to a person's identity and the limitations of this form of evidence.

4. Provide brief notes on three ways in which hair can provide forensic evidence.

5. What are the problems associated with measuring stature in both the living and the dead?

6. Explain why measuring the levels of lead-210 rather than carbon-14 could provide a more accurate estimation of the time elapsed since the death of a man who died in the late twentieth century.

7. Explain why it is easier to distinguish the difference in age between children of 4 and 10 years than between adults of 30 and 50 years on the basis of their teeth.

8. A young man has been found suspended from a dog leash. His neck has been broken and he has a note in his pocket saying 'Sorry Mum'. Why is this death suspicious?

9. Under what circumstances can a knife form a wound track longer than the blade?

10. Distinguish between the circumstances in which permanent and temporary cavities would be caused in firearm wounds.

11. Distinguish between the characteristics of a self-inflicted bite, a love bite and an aggressive bite.

12. What evidence would you look for to determine whether a person was alive or dead at the time a fire began?

13. State two means by which you could distinguish between a wound caused after death from one inflicted whilst the victim was alive?

Project work

Title

The influence of age, gender and environment on the longevity of fingerprints.

Rationale

It is stated in the literature that the fingerprints of children do not preserve as well as those of adults. It would be interesting to determine whether particular features of fingerprints are more susceptible than others and how environmental conditions such as temperature and humidity affected the results.

Method

Fingerprints of male and female volunteers of varying ages (preferably including babies to persons over 70) would be collected onto glass slides and strips of metal and paper. The volunteers would be required to have not washed their hands for an appropriate length of time beforehand – this would need to be determined in an initial series of experiments using different soaps. The slides and strips would then be stored in enclosed chambers at varying temperatures and humidities for different lengths of time before being developed and compared.

Title

DNA profiling from lip prints.

Rationale

There is relatively little information on the factors affecting the identification of DNA from lip prints.

Method

Lip prints would be made on various surfaces by drinking from a cup of hot tea, cold water or a glass of beer or by students kissing their own hand or that of another person. Students who are friendly with one another might kiss one another before making the print to determine whether DNA becomes transferred between individuals. The effect of lip salves and cosmetics on the isolation of DNA could be included in the analyses. The DNA would be extracted and sequenced using standard procedures. These experiments could be repeated using fingerprints as the source of DNA.

Title

The influence of biological decay on bullet markings.

Rationale

When a bullet is fired from a gun, it acquires a unique set of grooves that can be used to identify the gun that fired it. Decaying tissues and microbial action surround a bullet that remains in a dead body and this environment may affect the preservation of these identifying characters.

Method

Bullets fired from a single gun (the local gun club or territorial army brigade may be able to oblige) would be collected and placed in meat (e.g. minced beef or liver) that was then allowed to decay either above or below ground. This would therefore determine whether the presence of maggots influenced the results. The markings on the bullets would be compared at varying time intervals.

4 Bacteria and viruses in forensic science

Chapter outline

The role of microorganisms in the decomposition process

Microbial profiles as identification tools

How microbial infections can predispose people to crime and simulate the occurrence of a criminal act

The use of microorganisms in bioterrorism: anthrax; plague; smallpox

The transmission of HIV as a criminal act

The role of microbes in food poisoning

Objectives

Describe how microorganisms influence the decay process and the factors that affect the spread and growth of microorganisms in a dead body.

Discuss how microbial profiling techniques might be used to identify an individual or a location.

Discuss the possibility that an underlying infectious disease may predispose an individual to commit a crime.

Critically evaluate the risks posed by the criminal use of microorganisms and the techniques used to detect their presence.

Describe how the transmission of HIV can, in certain circumstances, be considered a criminal act and the mechanisms by which the perpetrator may be identified.

Introduction

The involvement of bacteria and viruses in legal cases is increasing as advances in technology, especially molecular biology, facilitate their identification, and courts of law become more willing to accept the reliability of DNA-based

Essential Forensic Biology, Alan Gunn
© 2006 John Wiley & Sons, Ltd

evidence. In addition, there is enhanced awareness amongst the public of the dangers posed by pathogenic bacteria and viruses being spread deliberately or through reckless behaviour by naturally infected individuals and this has led to changes in the law. Furthermore, there has been an increase in the numbers of individuals and groups threatening to release pathogens maliciously, or simply to cause distress and gain publicity. The overwhelming majority of such threats are hoaxes but the need for rapid pathogen identification and tracing has become a priority, as has the need to prosecute those making the hoax claims.

The role of microorganisms in the decomposition process

Many of the bacteria that are normally present upon or within our bodies are also extremely important in the decay process and anything that restricts their activity, such as low temperature, lack of oxygen or the absence of water, will prolong decomposition. In a very dry environment, especially if combined with ventilation and extreme heat or cold, a human body will mummify and can remain in this state for hundreds or even thousands of years. Similarly, under constantly frozen conditions a body, if undisturbed, will last for many years. Conversely, under warm, moist conditions, such as the tropics, a body will decompose extremely quickly even in the absence of invertebrates and other detritivores. There are few studies on the changes in the bacterial flora in and around a dead body (e.g. Hopkins *et al.*, 2000; Tibbett *et al.*, 2004) and, as yet, there are insufficient data to suggest how they could be used to determine the time since death. Bacterial colonization of a dead body usually begins from the intestine, which is naturally home to large numbers coliforms, bacilli and micrococci. As the body starts to decay, the pH becomes more acidic owing to the release of acids during autolysis (Chapter 1) and the products of bacterial fermentation, and it becomes anaerobic because all the oxygen is used up by the bacteria and the circulation system has ceased to operate. Consequently, the majority of the bacteria found in a dead body tend to be anaerobic and spore forming, such as *Clostridium perfringens*. Formerly known as *C. welchii*, this bacterium is found both within the guts and the female genital tract and surrounding skin. There are numerous strains of *C. perfringen*s and it is implicated in a range of pathogenic infections, ranging from food poisoning to cellulitis and gas gangrene. This is another good reason why care should be taken when handling dead bodies and tissue samples. Wounds, whether formed before or after death, are another source of entry and the bacteria spread rapidly via the blood vessels. Decay is therefore more rapid in a person who was suffering from septicaemia or other bacterial infections. By contrast, it is delayed where there is excessive blood loss because the lack of fluid in the blood vessels makes it harder for bacteria to penetrate the body.

Just as the tissues and chemicals that constitute a body are broken down by autolytic and microbial action after death, drugs present in those tissues may also become metabolized and degraded. The speed and extent to which break-down occurs depends upon the drug and the environmental circumstances. There is even the possibility that microbes may cause an increase in the drug concentrations. For example, gamma-hydroxybutyric acid (GHB) is naturally present at a very low level in our bodies but this level increases enormously when it is used as either a recreational or a therapeutic drug – or for criminal purposes. GHB is a class C drug that is used at low doses as an 'upper' by clubbers, who often know it as 'liquid ecstasy', but at higher doses it can cause confusion and even coma and is notorious for being used to spike drinks in 'date rape'. After GHB is ingested, there is an initial increase in concentration, after which the levels decline again within a few hours. There is also a rise in the levels of GHB following death, although not as high as those seen immediately after taking the chemical as a drug. It is therefore of forensic interest to know the extent to which levels of GHB found in a body might be ascribed to natural causes as opposed to misuse. *Pseudomonas aeruginosa*, a common bacterium associated with decomposing tissues, is capable of producing GHB although not in sufficient quantities to account for all of the natural increase, so there may be other bacteria and/or physiological processes contributing to the rise (Elliott *et al.*, 2004).

Microbial profiles as identification tools

The study of microbes has always been hampered by their small size and the difficulty of growing many of them in the laboratory. However, recent advances in molecular phylogenetic techniques have indicated that the microbial flora is actually far more diverse than the plant or animal kingdoms. Similarly, because they are so small, their distribution was considered to be less restricted than that of plants and animals and consequently they were not thought to abide by the same taxa–area relationship. This is the ecological principle that underlies the finding that larger islands tend to have more diverse fauna and flora. However, work by Green *et al.* (2004) and Horner – Devine *et al.* (2004) has indicated that this is not the case and microbial communities are not random. This opens up the possibility that it may be possible to use the microbial composition of soil and other debris found on a suspect as a potential identifying characteristic linking them with the scene of a crime. Although many microbes cannot be cultured *in vitro* it is possible to use molecular techniques such as 16S rDNA sequencing and automated ribosomal RNA intergenic spacer analysis (ARISA) to obtain a community 'genetic fingerprint' linking them to a specific geographical area. A feasibility study by Horswell *et al.* (2002) indicates that soil microbiological profiling has real forensic potential but more work needs to be done to optimize the methodology and determine its strengths and limitations.

Human bites can be nasty because they contain bacteria (Figure 4.1) capable of causing a serious wound infection. Clenched fist injuries, a frequent conse-

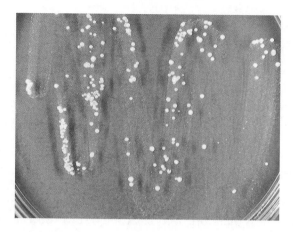

Figure 4.1 Bacteria grown from human saliva on nutrient agar. Numerous bacteria are growing despite the swab being taken from a person who had brushed his teeth and used a mouthwash only 2 hours previously

quence of hitting someone in the mouth during a fight, are particularly liable to become infected (Monteiro, 1995). However, these bacteria may be useful in identifying the person who inflicted the bite. The bacterial flora, especially the streptococci naturally present in our mouths, is extremely diverse and their genomic profile appears to be characteristic for an individual. Work by Borgula *et al.* (2003) and Rahimi *et al.* (2005), in which volunteers were asked to bite themselves after which samples were taken over varying time periods, demonstrated that live bacteria could be recovered from the bite site for at least 24 hours provided that it was relatively undisturbed, as well as from fabrics. Furthermore, the genomic profiles of the bacteria recovered from the bite sites provided a unique identifier of the person responsible. Obviously there is a great deal of work to be done to confirm the reliability and limitations of this technique and to agree a standardized protocol, but it could prove to be a valuable tool in forensic analysis in the future.

The possibility that humans might be identified individually by differences in Epstein-Barr virus polymorphisms has been suggested by Kobayashi *et al.* (1993) (www.promega.com/geneticidproc/ussymp7proc/ab42.html), although their work has yet to be substantiated in peer-reviewed publications. According to Kobayashi and his co-workers, because many humans are infected with the Epstein-Barr virus (EBV), and the virus is genetically highly polymorphic in short tandem repeat sequences, it should be possible to detect this variation using techniques such as PCR and Southern blot analysis of human tissue samples – although as the virus is found in white blood cells, blood samples would probably be better. They claim that this approach has enabled them to distinguish between 'identical twins with exactly the same DNA'. Children living in low socio-economic groups tend to become infected with EBV earlier in life than those from affluent families but by adulthood 80–90 per cent of the population tends to exhibit serological evidence of past infection, although there are geographical variations. EBV is usually spread via saliva and may be asymptomatic or give rise to serious diseases such as infectious mononucleosis (glandular fever) and Burkitt's lymphoma. Ikegaya and Iwase (2004) have demonstrated that in Japan it is possible to trace the origins of unidentified bodies by genotyping the JC virus (JCV). JCV is a polyoma virus that infects most humans during childhood and is usually asymptomatic, except in immuno-compromised people. Once someone is infected, the same viral strain can be found in their kidneys for the rest of their life. In addition to being found in the tissues, the virus is also excreted in the urine – this may have forensic impor-tance where urine splashes are found at a crime scene. However, polyoma viruses are very delicate and difficult to grow in culture so the urine splashes would probably have to be fresh. It is possible to extract JCV DNA from bodies that are up to 10 days old, hence facilitating the identification of corpses that are already badly decayed (Ikegaya *et al.*, 2002). The forensic potential of JCV geno-typing has not yet been assessed in other countries and would not allow a positive identification of an individual's identity but, in conjunction with other evidence, has the potential of suggesting the area where the unidentified person grew up, which would help speed up the investigation.

How microbial infections can predispose people to crime and simulate the occurrence of a criminal act

Many infectious agents can bring about changes in behaviour that can lead pre-viously well-behaved persons to commit crimes. Syphilis, for example, is well known for causing dementia during the later stages of infection. Lyme disease, caused by the spirochaete *Borrelia burgdorferi*, which has become a widespread problem in the USA and many parts of Europe, has now also been added to the list of bacterial infections causing behavioural changes. Previously thought to be a predominantly arthritic illness, it is now known to cause mental dis-orders (Hassett *et al.*, 2003; Rudnik *et al.*, 2003) and this has already been used as a defence by a man accused of attacking a stranger with an axe for no known reason (http://216.117.159.91/crimetimes/99b/w99bp11.htm). Such infections create medico-legal problems because in many, such as Lyme disease, only a small proportion of those suffering the disease may suffer mental prob-lems, and of those a smaller percentage will commit offences as a result. It is therefore a difficult clinical judgement to determine whether or not a person committed a crime as a consequence of being infected with the disease. In a similar vein, the recently described condition PANDAS (paediatric autoimmune neuropsychiatric disorders associated with streptococcal infection) has been linked, amongst other things, with obsessive–compulsive behaviour. This in turn has lead to suggestions that prisoners, many of whom have a variety of mental illnesses, might benefit from being treated with antibiotics. However, according to a review by Kurlan and Kaplan (2004), the scientific evidence that PANDAS really exists as a clinical entity is not yet sufficiently proven.

Some microbial infections result in the formation of serious skin lesions that might be mistaken for signs of physical assault. This is especially the case in vulnerable persons, such as the very young, the mentally disturbed and those suffering from senility, who are unable to relate what has happened to them. For example, Nields and Kessler (1998) describe a case in which a 4-year-old child who died of streptococcal toxic shock syndrome was initially thought to have been a victim of child abuse. Fortunately, the bacteria were identified from samples taken at autopsy and no-one was charged with assault. Streptococcal toxic shock syndrome is caused by infection with any group A streptococcal bacteria and may result in organ failure and shock, the consequences of which can be rapidly fatal. In many cases it is not known how the bacteria enter the body but a common feature of the syndrome is the development of necrotizing fasciitis in which there is destruction of the fascia (fibrous tissues that cover and separate the muscles) and fat. Following this, the overlying skin may die, split and break and the damaged region may, at first sight, be thought to have resulted from a wound that has become infected.

It is essential to exclude the possibility that an infectious agent might be involved in an otherwise 'unexplained' death. This message was brought painfully home in the recent unfortunate case of Sally Clark. Mrs Clark was tried and convicted in 1999 of murdering two of her infant sons. The first son, 11-week-old Christopher, died suddenly in 1996 and 14 months later her second son, 8-week-old Harry, also died. The medical profession therefore became suspicious and a police investigation ended with Mrs Clark being sentenced to life in prison. The same pathologist carried out autopsies on both children. Initially he ascribed Christopher's death to a respiratory tract infection but after Harry died he revised his opinion and suggested that he had been smothered. Harry's death he ascribed to 'shaken baby syndrome'. A second pathologist also concurred that the deaths were suspicious. However, it subsequently transpired that the first pathologist had withheld the results of microbiological tests carried out as a normal part of the autopsy routine. These demonstrated that Harry was suffering from an active infection with *Staphylococcus aureus*, the bacteria being found in eight regions of his body, including the cerebrospinal fluid. *Staphylococcus aureus* is commonly found on the surface of the skin and in the nose of normal healthy individuals and is generally considered harmless to such people – because of this, the pathologist did not submit the findings to the coroner because he considered the bacteria to be a consequence of contamination. However, *S. aureus* is also associated with skin infections such as boils and abscesses and it is capable of causing potentially fatal septicaemia and meningitis. The presence of the bacteria in Harry's body, and especially in the spinal fluid, together with his blood cell profile, indicated that Harry could have died as a result of meningitis rather than a criminal act. This resulted in Mrs Clark being freed on 29 January 2003 and her conviction being declared 'unsafe' (Byard, 2004; Dyer, 2005).

The use of microorganisms in bioterrorism

Biological warfare has been practiced in a minor and infrequent way for hundreds of years but its impact has been small compared to conventional weaponry. The development of biological warfare agents was banned under the Geneva Convention many years ago – not that this put a stop to it! During the Cold War era, both Western governments and Eastern communist regimes devoted considerable effort to the development of agents causing human and animal diseases as a means of causing huge casualties to the enemy. Following the physical and economic collapse of the Soviet Empire there has been inter-national concern that the diseases developed by the Soviet scientists and their knowledge of how to produce and deliver them might find their way into the hands of terrorists and 'rogue governments' who bear grudges against the Western world (Brumfiel, 2003). The USA is so convinced of the threats that it is currently spending billions of dollars in developing strategies to combat them. With so much money readily available it is not surprising that some scientists are talking up the threat in their search for research funds and advancement. This is not to deny that very real risks exist where state-sponsored production of biological weapons is concerned but there is also a great deal of hype, which is further fuelled by a media in constant search of disaster stories. The US Centres for Disease Control and Prevention (CDC) have categorized pathogens in a list (A–C) according to their potential for use in bioterrorism, with those in List A considered to be the most dangerous. The UK operates a similar clas-sification system in which organisms are graded 1–4, with those in grade 4 being considered the most dangerous.

Biological warfare agents are sometimes said to be 'cheap and easy' to produce, as if all microbes are the same, and that they are 'simple to deliver' to their target. Growing highly infectious, rapidly lethal bacteria or viruses usually requires Category 4 containment facilities and highly trained scientists other-wise the production staff would soon become infected and die. Even with access to the facilities of a modern biodefence laboratory, working with dangerous pathogens is a risky occupation. For example, in 2004 three researchers at Boston University's medical campus working on the tularaemia bacterium became infected (Dalton, 2005). Fortunately, the infections were not fatal and it is not normal for tularaemia to be spread by person-to-person contact. However, it was some time before it was realized that these infections had occurred and with a different pathogen it would have been possible that the disease could have unintentionally spread into the nearby community. Ter-rorists would also face the further problems of establishing a suitable dose to incapacitate or kill the intended target(s) and a means of delivering that dose.

Biological agents cannot be delivered in a conventional bomb because the microbes would be destroyed in the heat of the explosion. It has been suggested that a disease might be spread by a suicidal 'biobomber' becoming intentionally infected and then infecting the target population by travelling around a city on crowded public transport. Dying in a sudden explosive 'blaze of glory' is one thing but intentionally dying over a period of days from a painful incapacitating disease, in a public place and without drawing attention to oneself, calls for a different and very rare type of suicidal individual. The use of pathogens by bioterrorists is therefore most likely to be aimed at causing terror and disruption of public services rather than widespread mortality. The release of the agent would probably be publicized by the terrorists at the time of release, just as is the case with many of their conventional bombs, or there would be the threat of a release in order to blackmail the government – this would require very small amounts of agent and the delivery device need not even be very effective but would still result in widespread publicity and panic. These issues are, however, beyond the scope of this section.

It should be remembered that the majority of cases of infectious diseases being spread through the population are a consequence of natural outbreaks or human negligence. For example, in August 2002 almost 200 people at the Barrow Arts Centre were infected with legionnaires' disease (caused by the bacterium *Legionella pneumophila*), of whom seven died. This occurred as a consequence of the bacteria being allowed to replicate within an air-conditioning unit that then sprayed them into the surrounding air. This has led to charges of unlawful killing against the technical and design services manager responsible for the site and the local council (*The Independent*, 9 February 2005). In a case of suspected bioterrorism, forensic science becomes involved following the identification of a pathogen and the consequent need to determine where it originated (Breeze *et al.*, 2005). This might be done from a comparison of its genetic profile or stable isotope ratio (Kreutzer-Martin *et al.*, 2004) with that of cultures held legitimately by laboratories throughout the world. Ideally there should be a list of all persons who have access to the cultures in each laboratory. This requires a reliable database and a level of cooperation between countries that is steadily improving but far from perfect. For example, from the molecular characteristics of anthrax bacilli it would be possible to determine whether a person was suffering from a naturally acquired infection or from a variety that had been modified in the laboratory. However, for a number of possible biological agents the tests are not available to enable identification to strain or sub-strain level and/or the tests have not yet been fully validated. A wide range of viruses and bacteria have been suggested for possible use in biological warfare (see following websites for more details: www.prodigy.nhs.uk; http://omni.ac.uk/browse/mesh/C0872021L1713313.html) and Table 4.1 features those that are most likely. A few examples are considered here to illustrate how they have been used in the past and how their source has been identified.

Table 4.1 Summary of the human pathogenic microorganisms most likely to be exploited as bioterrorist weapons

Microorganism	Mode of transmission
Viruses	
Smallpox	Inhalation
Hantavirus pulmonary syndrome (HPS)	Inhalation
West Nile virus	Mosquito bite
Ebola virus	Contact
Marburg virus	Contact, sexual transmission possible
Lassa fever virus	Contact, sexual transmission possible
Bacteria	
Anthrax, *Bacillus anthracis*	Inhalation, ingestion
Plague, *Yersinia pestis*	Flea bite, inhalation (pneumonic plague)
Salmonella food poisoning, *Salmonella typhimurium*	Ingestion
Typhoid, *Salmonella typhi*	Ingestion
Cholera, *Vibrio cholerae*	Ingestion
Tularaemia, *Francisella tularensis*	Tick bites, ingestion, contact
Brucellosis, *Brucella* spp.	Contact, ingestion
Botulism, *Clostridium botulinum*	Ingestion

Anthrax

Anthrax is caused by the bacterium *Bacillus anthracis*, which owes much of its pathogenicity to the secretion of a toxin containing three proteins. One of these, protective antigen, facilitates the entry of two toxic enzymes, lethal factor and oedema factor, into the host cell. The genes coding for the toxin are found on plasmids and these have now been sequenced genetically. In addition, the full genomic sequence of several strains of *B. anthracis* is now available, which will facilitate diagnosis and improve our understanding of how it causes disease. The bacteria are relatively large and rod-shaped; in clinical specimens the bacteria have a capsule and are usually seen singly or in twos, and spores are not present. By contrast, in culture anthrax bacilli form long chains, they do not have a capsule and spores are produced. Anthrax spores are notoriously difficult to destroy and they can survive in soil for many years. In 2004 a batch of anthrax, which was shipped to researchers at the Children's Hospital and Research Centre at Oakland, USA, for vaccine production was found to contain live and still lethal bacteria despite having been heat inactivated and tests being done to check that they no longer grew in culture (Anon, 2004b). The ability of the spores to survive for long periods of time coupled with the possibility of engineering the bacteria to express enhanced pathogenicity and resistance to anti-biotics has led the CDC to categorize anthrax as a List A organism.

Clinically, there are three principal forms of anthrax: cutaneous, gastroin-testinal and inhalation. Persons suffering from cutaneous anthrax present with a painless papule that subsequently develops into a black necrotic ulcer; this is followed by an extensive red oedema along with a high fever and bacteraemia – death ensues if there is no treatment. This is the most common form of human disease and is naturally acquired through breaks in the skin by persons who work with infected animals or animal products such as leather. It used to be a relatively common disease and was known as 'woolpackers' disease' in the UK from its association with the trade. Following improvements in animal care, hygiene and the availability of an effective animal vaccine, the disease is seldom seen in humans in developed countries. In the UK there were only 14 cases between 1981 and 2000 (www.dh.gov.uk), whereas in the USA there was an average of 5 cases a year between 1955 and 1999. Gastrointestinal anthrax is acquired through eating food contaminated with anthrax spores, which causes lesions in the gastrointestinal tract through which the bacteria invade the rest of the body, causing septicaemia and death. Naturally acquired cases of inhalation anthrax are very rare. It results from breathing in the spores and is therefore the form most likely to result from a bioterrorist attack. Initial symptoms resemble those of a common cold but once the lungs are infected severe breathing difficulties ensue and the disease is rapidly fatal. Human-to-human transmission of anthrax is not thought to occur.

Different strains of anthrax differ markedly in their pathogenicity, a trait used both in the development of vaccines where a lack of pathogenicity is a virtue and the development of biological warfare agents where high pathogenicity is required. The Aum Shrinikyo cult in Japan were probably not aware of this when they released an aerosol of anthrax spores in Kameido (which is situated near Tokyo) in 1993. Cult members sprayed a liquid suspension of spores from the eighth floor of their headquarters building but nobody was aware of this at the time, although several people complained of the smell. The act only came to light later following the cult's more successful release of Sarin gas on a Tokyo underground station, which resulted in several deaths and a thorough investi-gation of the cult's activities. Retrospective analysis of fluid samples collected and stored by the authorities as a result of the complaints about the smell resulted in the discovery of anthrax spores from which it was possible to grow colonies of bacteria. DNA isolation and MLVA (Multiple Locus Variable number tandem repeats Analysis) genotyping of these colonies demonstrated that the anthrax belonged to the Sterne vaccine strain (Keim et al., 2001). The Sterne34F2 vaccine strain is widely available in Japan where it is used in the preparation of animal vaccines and it is therefore probable that a cult member obtained a vial of animal vaccine from which a large number of bacteria were subsequently cultured and released. Thankfully, being non-pathogenic, it posed little risk to humans.

Such analyses are not always as simple and straightforward, and despite all the efforts put into sequencing the anthrax genome investigators are still no nearer to identifying the source and culprit responsible for the anthrax attacks in USA in 2001. During October of that year a number of letters containing anthrax spores were sent through the US mail to media outlets and two US senators. This resulted in 18 confirmed cases of anthrax from which 5 people died (www.niaid.nih.gov/factsheets/anthrax.htm). Of these cases, 7 people contracted cutaneous anthrax, all of whom survived, and 11 contracted inhalation anthrax of whom only 6 survived. This indicates the higher incidence of inhalation anthrax when contracted via a bioterrorism attack and the greater difficulty in treating this form of the disease. It also demonstrates how difficult it can be to identify the source of a pathogen in the absence of any claims for responsibility. It also indicates that whoever carried out the attack probably had access to limited quantities of the bacteria because it remains the only confirmed case of anthrax being transmitted via the mail in the USA. He/she also had access to equipment necessary to grow the bacteria, place them in letters, post them and still not become infected him/herself. There is a strong suspicion that the person responsible acted alone and was at some point connected to a US military / medical research institute.

As a consequence of this case, a climate of heightened awareness has developed in many countries. Anthrax spores do not even have to be present, or even threatened to be present, to cause panic and disruption. On 16 October 2001, Liverpool's Royal Mail sorting office had to be evacuated and workers passed through a mobile decontamination unit when a fine powder was seen leaking from an envelope – this subsequently proved to be sand from Australia. This illustrates the need for rapid identification methods for pathogens that do not rely on them being first grown in culture. Liquid chromatography–mass spectrometry (LC–MS) is one of several techniques showing potential in this area. Bacterial cells are first disrupted in an ultrasonicator to release their proteins, which are then filtered and separated on a multidimensional capillary LC (liquid chromatography) system. Proteins that have previously been shown to be reliable biomarkers can then be detected using a mass spectrometer. A total analysis time of approximately 25 minutes has been claimed and a sensitivity in the order of 10^{-15} moles of biomarker protein (De Boek *et al.*, 2002).

Plague

The devastating outbreaks of bubonic plague that afflicted the UK and other European countries during the middle ages have become irrevocably burnt onto the national psyche and the mere mention of 'plague' is sufficient to cause anxiety amongst the population. Because the last serious outbreaks in the UK took place over 200 years ago, many people erroneously believe that plague now exists only in the mists of history, although it is still recorded in various parts of Asia and there is a natural cycle between prairie dogs in North America from which humans are occasionally infected. Although there have been suggestions that the great plagues of Black Death in the Middle Ages were due to an unidentified virus related to the Ebola and Marburg viruses (Scott and Duncan, 2004), the majority of microbiologists continue to believe that the bacterium *Yersinia pestis* was responsible. Plague exists in three principal forms, bubonic, pneumonic and septicaemic, all of which are caused by *Y. pestis*. Bubonic plague is characterized by the formation of painful swellings (buboes) of the lymph nodes (Figure 4.2), principally the groin, axillary and cervical nodes, and may subsequently develop into pneumonic and/or septicaemic plague. Provided that it is treated early enough, bubonic plague has a low fatality rate of about 8 per cent. Pneumonic plague may result directly from droplet infection or develop from bubonic plague and is rapidly fatal. Septicaemic plague may develop without any previous signs of disease or from the bubonic / pneumonic forms. It results in septic shock and has a high fatality rate of 33 per cent or more. Plague has the distinction of being involved in the first documented use of biological warfare when the Mongol armies encamped around the Black Sea port of Kaffa catapulted the bodies of men who had died of plague over the city's battlement walls. However, the effectiveness of this is questionable because plague is usually transmitted by fleas or by droplet infection in the case of the less common pneumonic form. Fleas leave the body when it starts to cool and would probably have gone before the body was sent airborne, and

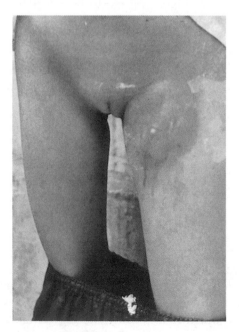

Figure 4.2 Swelling of the lymph glands to form buboes is a characteristic symptom of bubonic plague. Sometimes, as in the case illustrated here, the swellings will burst and suppurate. (Reproduced from Peters and Gilles, 1977, with permission from Wolfe Medical Publications)

droplet infection from a corpse is unlikely. Contaminated clothing is thought to present a low risk of infection (www.prodigy.nhs.uk/guidance.asp?gt= Plague%20-%20deliberate%20release). *Yersinia pestis* is currently one of the microbes considered to pose a risk as a bioterrorist weapon and the CDC classifies it as a List A organism. Owing to changes in living conditions, the cycling of disease between rats (the normal host of the bacteria) and humans would be difficult to establish in developed countries, so releasing infected rats (which would also need to be carrying the rat flea *Xenopsylla cheopsis* to transmit the infection) would be unlikely to result in more than the death of the rats. The disease would therefore have to be released in droplet form to be effective and be in sufficient concentration for a person to breathe in 100–500 bacilli. Owing to the rapidity with which pneumonic plague kills and how ill someone with pneumonic plague would be, it is extremely unlikely that a suicidal biobomber would be physically capable of walking around infecting the population. The disease would therefore have to be released from some sort of vapour-producing device within an enclosed space. Like all terrorist weapons, however, the threat far outweighs the potential to cause deaths. A natural outbreak of bubonic and pneumonic plague in Surat, India, in September 1994 resulted in widespread panic and the city being officially closed. Many people fled the area – something guaranteed to spread the infection.

Smallpox

Smallpox is caused by a Variola virus, and is related to cowpox virus and monkeypox virus; chickenpox virus is a herpesvirus and is not related to smallpox. The term 'pox' is an old word and refers to a disease that causes the formation of pustules. Smallpox gained its common name not in reference to the size of the pustules but to distinguish it from syphilis, which was referred to as the 'great pox'. It is spread by droplet infection, by direct contact with an infectious person or from contact with clothing or bedding previously used by a person suffering from smallpox. The incubation time is approximately 3 weeks, after which flu-like symptoms are expressed; a person becomes infectious 2–3 days before these symptoms appear. A few days after the onset of the disease, the patient's temperature returns to normal and he/she develops a vesicular rash on the face and limbs that spreads to cover the whole body (Figure 4.3). Lesions in the mouth and throat release large amounts of virus that are then aspirated in the saliva and droplets in the breath. Over the course of 7–14 days, these vesicles enlarge, rupture and start to heal over, at which point the patient ceases to be infectious. However, in about half of all cases the fever then returns and widespread internal haemorrhage leads to death within a few days. The disfiguration and high mortality make smallpox a very frightening disease but it is also a rare example of how a disease can be eradicated – the world was officially declared free of smallpox in 1979. Eradication was possible because an effective vaccine was available, everyone who is infected with the virus devel-

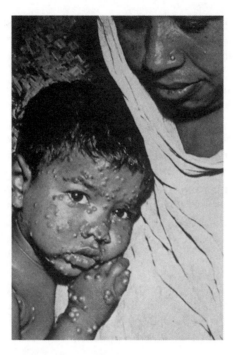

Figure 4.3 Child suffering from smallpox. The skin lesions can become secondarily infected with bacteria; pneumonia and bacterial septicaemia are frequent complications. Although the lesions are fairly characteristic, confirmation of the disease depends upon laboratory identification by serology, PCR or electron microscopy. (Reproduced from Peters and Gilles, 1977, with permission from Wolfe Medical Publications)

ops the symptoms (they could therefore be identified and treated before the disease was transmitted further) and all countries in the world were willing to cooperate with the World Health Organization eradication campaign.

Although the disease was eradicated, stocks of the virus were maintained in the USA at CDC Atlanta and in Russia at the Vektor Institute, Novoskibirsk, ostensibly so that vaccines could be developed should the disease re-emerge. There have been continuous rumours that the security of these stocks may have been breached and that other sources are held elsewhere in the world. The countries and organizations that are accused of holding illicit stocks all deny this but it is impossible to prove a negative and the consequences if the disease was released are dire. For example, in the UK, vaccination against smallpox ceased in the 1960s and therefore by 2005 anyone under 40 years old would not have been vaccinated and the immunity of older people would have declined – therefore the majority of the population would be susceptible. Computer models of how the disease would spread indicate that within 3 weeks ten people deliberately infected from an initial 'index case' (e.g. a 'biobomber') would each have inadvertently infected a further ten people (i.e. a total of 100 people). It should be remembered from the symptoms listed above that the 'biobomber' would have a very short 'window of opportunity' during which he would be infectious but his diseased condition would not be patently obvious from the facial rash and he would probably be feeling very ill. Regardless of how the virus was introduced into the population, the diagnosis of smallpox anywhere in the world would be an indication of the release of illicitly held virus. The UK government has drawn up contingency plans for dealing with smallpox, as it has for all biological and chemical weapons (www.dh.gov.uk; www.hpa.org.uk), in which it is expected that the virus would spread rapidly owing to the low immunity in the population, the speed and frequency with which people travel during their daily lives and the unfamiliarity of people, including doctors and nurses, with the symptoms of the disease. Alert levels (1–6) are to be declared by the Chief Medical Officer and if there is 'an overt release' (i.e. the terrorists gave a specific warning) the contingency plan would be led by the police. If the release is 'covert' (i.e. there was no warning but cases of disease were positively identified), the plan would be led by the Department of Health. Regional smallpox diagnosis and response groups have been established, which include a smallpox diagnostic expert and smallpox management and response teams. Confirmation of the disease would require electron microscopy and PCR analysis of the virus. At Alert Level 3 (Outbreak occurring in the UK), the contingency plan is to control the spread of smallpox and minimize disruption and inconvenience by identifying positive cases and isolating them in 'smallpox centres', then identifying all their possible contacts and vaccinating them. Mass vaccination of the whole population would only be considered at Alert Level 4, when there would be large multiple outbreaks across the country in which the cases could not be linked. The value of vaccinating the general population has to be balanced against the one in a million chance of death resulting from reactions against the vaccine itself (in a population of greater than 60 million, this is a serious consideration) and the problems of supplying sufficient doses of vaccine.

The transmission of HIV as a criminal act

Human immunodeficiency virus (HIV) is a human retrovirus that affects the immune system and in time causes the condition 'acquired immunodeficiency syndrome' or AIDS. A person may remain HIV positive but apparently healthy for months or even years before symptoms of AIDS become apparent. The virus infects and destroys cells bearing the CD4 molecule, which includes T-helper cells and a subset of blood and tissue macrophages. Following replication, the pro-viral DNA genome becomes integrated into the host chromosome where it may remain quiescent or enter a cycle of production of progeny virus that results in the cell's destruction. The clinical manifestations of HIV infection are a result of depletion of the T-helper cell population and thereby an impairment of the body's ability to respond to other infectious agents.

HIV/AIDS is a major problem throughout the world and health workers have to treat all patient samples as though they might be infected. In the UK it is illegal to test a person for HIV without their consent because they must have counselling on being told the result – so unless a patient or person in police custody already knows that they are HIV positive and volunteers the information, or agrees to a test, there is no way of finding out. Forensic pathologists are well aware of the risks of contracting HIV and other diseases, such as hepatitis B (HBV) and hepatitis C (HCV), whilst performing an autopsy and that this risk does not disappear even if the body is many days old (Douceron *et al.*, 1993; du Plessis *et al.*, 1999; Nolte and Yoon, 2003). Consequently, anyone handling a dead body or body parts should wear cut-resistant undergloves, appropriate protective clothing and work with extreme caution (Galloway and Snodgrass, 1998). In a similar manner, the possibility of accidental transmission of HIV during sports such as rugby, karate and boxing, where there is forceful physical contact between the participants, has become a source of concern, although the chances of this occurring on a playing field or sports hall are considered to be extremely low. In 1989 an Italian footballer reportedly contracted HIV following a collision on the field with an infected player, although it was not conclusively proven that this was the source of the infection.

Although there are many documented cases of health workers contracting HIV during their work, usually through needle-stick injuries, there are few reports of patients contracting the disease from their carers. A very prominent case where this did occur was that of David Acer, a Florida dentist, who, following diagnosis of being HIV positive, continued to practice without telling his patients. Even more importantly, he used poor infection control practices. For example, it was alleged that he did not always sterilize his instruments or change his gloves between patients. Several of his patients contracted HIV despite being in low-risk groups and subsequent genetic analysis indicated that the HIV strains that both they and David Acer carried were all closely related. Furthermore, other HIV-positive people living in the vicinity of Acer's practice had different strains of the virus, which suggests that his patients had contracted their infection from him rather than their neighbourhood (Ciesilski *et al.*, 1994). This remains an isolated case and it is still uncertain how exactly transmission from dentist to patient was effected. In a subsequent study of another HIV-positive dentist in Florida, who also admitted that he did not always follow the full infection control guidelines, there was no proof that he transmitted his infection to his patients (Jaffe *et al.*, 1994). Although 28 of his patients were HIV positive, 24 of them were considered to be at risk owing to their lifestyle, and genetic analysis failed to indicate a link between the viral strains with which the dentist and his patients were suffering.

Following a change in the UK law, persons who know that they are HIV positive and do not inform their partner(s) before indulging in unprotected sex, or bite or otherwise intentionally attempt to infect someone, are now liable to prosecution and can be sent to jail. Criminalizing the sexual transmission of HIV has proved extremely controversial, as is anything connected to HIV, its treatment and transmission. One of the arguments against it is that it might prevent someone from seeking an HIV test because if you do not know that you are infected with a disease you cannot be accused of knowingly spreading it. From a strictly legal point of view, proof of malicious intent is difficult because, where sex was consensual, it relies on one person's word against another's that he/she intended to commit harm. By comparison, proving the source of the virus is a relatively easy matter of analysing the genetic profile of the virus isolated from the defendant and the plaintiff. Successful prosecutions have, however, been brought. In a landmark UK case in March 2001, a man who had known of his HIV status for almost a year was prosecuted for allowing his girlfriend to become infected. His girlfriend claimed that she was not aware of his HIV status when they began their sexual relationship and he only admitted it to her after she had discovered from a blood test that she too had become HIV positive. The man denied this version of the events but the jury did not believe him and he was jailed for 5 years. The chances of transmission of HIV and other sexually transmitted diseases during rape or sexual assault are much higher than in consensual sex owing to the violence involved. Victims of assault are therefore always offered an HIV test and counselling when they report the attack. In parts of Africa, a rumour has spread that sex with a virgin is a cure for the disease and this has led to a terrible increase in the number of rapes of young girls.

The threat of transmitting HIV is being increasingly used by persons resisting arrest, either by spitting at the police officers or attempting to bite them. For example, in January 2004, a man in London became drunk and disorderly a month after finding that he was HIV positive. In a struggle with police attempting to arrest him, he bit one officer and attempted to bite the face of another whilst shouting 'I will bite your face and give you AIDS' – he was not successful in this and in court he pleaded guilty to common assault and using threatening behaviour and was jailed for 4 months. As a consequence of cases such as this, the police take spitting extremely seriously and are also attempting to change the law so that they can test subjects for HIV, HBV and HCV without their consent. A remarkably heartless case of deliberate HIV transmission was heard in St Charles, Missouri, USA in 1998 in which a father, who worked as a phlebotomist, stole infected blood from his place of work and injected it into his 11-month-old son who was lying sick in hospital suffering from an asthma attack. Doctors treating the child became concerned when a blood test revealed that he had somehow contracted HIV. They contacted the police, who became suspicious of the father when he reportedly told his ex-girlfriend that she would never claim any child support off him because the boy would not live that long. The father was convicted on the basis of largely circumstantial evidence and received a life sentence; the child went on to develop AIDS.

In Libya, an unusual case has arisen in which five Bulgarian nurses and two doctors, one Bulgarian and one Palestinian, have been sentenced to death for intentionally infecting about 400 children with HIV, 40 of whom have already died. The verdict has caused an international outcry, not least because of over-whelming forensic evidence indicating that they are innocent (Anon, 2004c). Eminent French and Italian viral experts employed by the Libyan government to investigate the case demonstrated that the children were infected before the Bulgarian nurses arrived in Libya and the high incidence of co-infection with HBV and HCV implied that the infections were acquired through poor hygiene practices at the hospital rather than a deliberate act. Furthermore, genetic analysis showed that viral samples from the children were all very similar, suggesting a common origin – which would also point towards unintentional contamination from an already naturally infected patient.

The role of microbes in food poisoning

Food poisoning is extremely common and many people automatically assume that any bout of sickness or diarrhoea is a consequence of something they ate. However, some experts believe that up to 50 per cent of such cases are actually the result of infections acquired from other sources, such as pets, contact with faeces, contact with someone who was already infected or failure to follow simple hygiene rules such as washing hands (www.food.gov.uk). The most common causes of food poisoning in the UK are the bacteria *Campylobacter* spp. (which causes the majority of cases), *Salmonella* spp., *Escherichia coli* O157, *Clostridium perfringens* and *Listeria monocytogenes*. A number of viruses also cause food poisoning, such as Norovirus (Norwalk-like virus), rotavirus and hepatitis A. Most cases of food poisoning, although unpleasant, are not life threatening and occur as individual or sporadic events. An outbreak is defined as occurring when two or more people fall ill having consumed a common batch of food. Occasionally food poisoning may prove fatal, especially in the very young, the elderly and the infirm. Food poisoning usually results unintentionally from the incorrect storage or cooking of food, which allows the bacteria within it to survive and replicate to dangerously high levels, or from poor hygiene prac-tices, which result in the transfer of bacteria from contaminated food or surfaces to previously uncontaminated food. Identifying the source of an outbreak of food poisoning requires the offending organism(s) to be cultured from the patients' faeces and questioning the patients to determine their common food source. This may prove difficult because not everyone who consumes contaminated food need fall ill and if the food has all been consumed or thrown away it may be impossible to prove that it was the source of the infection.

The intentional contamination of food with foreign bodies (e.g. razor blades, glass) and poisons (e.g. mercury) has been common practice for many years, usually by disaffected individuals with a grudge against the manufacturer or society at large or with a view to blackmail. For example, in September 2003, 222 children were hospitalized in Hunan province, China, after eating food con-taminated with rat poison. The relative ease with which this can be done often spawns copycat attacks, and these have been a particular problem in Japan where there have been several fatalities.

There are few reports of the deliberate use of microbes, either through introducing them or through poor cooking or storage practices, in order to cause food poisoning. However, it is a possibility where someone wishes to cause distress but not death and would be difficult to prove, although it would carry a real risk of unintentionally causing fatalities. The US-based Rajneshee cult are known to have manually infected food in salad bars with *Salmonella typhimurium* (a common cause of diarrhoea) in an attempt to influence a local Oregon election by making people too sick to vote. Cult members obtained the bacteria via their own state-licensed medical laboratory, which in turn purchased them from commercial sources. They coated plastic gloves with the bacteria and then handled the contents of salad bars in eight restaurants, thereby causing illness in several hundred people. The cult's involvement only became apparent when it started to splinter and some members turned informant. It was alleged that the cult had considered using the much more deadly *Salmonella typhi*, the causative agent of typhoid fever, but this idea was abandoned because it would have been much easier to trace the source of the infection back to the cult. There are concerns that the extremely dangerous bacterium *Clostridium botulinum* might be used by bioterrorists, either through contaminating food or releasing its lethal toxin in the form of an aerosol into the atmosphere (www.prodigy.nhs.uk). The release of the bacteria or its toxin into the water supply is not thought to represent a major threat because existing water treatment and purification procedures are sufficient to remove them.

Quick quiz

1. Why does extensive blood loss slow down the rate of decay?

2. Why are the bacteria involved in decay predominantly anaerobic?

3. How might oral bacteria prove useful in identifying crime suspects?

4. Give an example of how a defendant might plead 'suffering an underlying infectious disease' as a mitigating circumstance for committing a crime.

5. Why do anthrax spores differ in their pathogenicity?

6. What are the three principal forms of plague?

7. Why would even one case of smallpox anywhere in the world be evidence of illicit activity on behalf of a terrorist or government organization?

8. How would it be possible to identify a person recklessly spreading HIV infections?

Project work

Title

Soil microbial profiles as forensic indicators.

Rationale

The soil microbial flora is extremely diverse but its usefulness as a forensic indicator linking a person, animal or object to a location is not known.

Method

The microbial profiles would be determined by procedures such as 16S rDNA sequencing and automated ribosomal RNA intergenic spacer analysis (ARISA). Soil samples would be taken from the clothing and the soles of shoes and the microbial profiles compared to those obtained from the location where the person had been walking. They would also be compared to other locations in the area and the effect of storing the samples on the microbial profile would also be assessed. For example, does the microbial profile change if the sample is not analysed for two or more days?

Title

The effect of a dead body on the underlying soil microbial flora.

Rationale

It is sometimes useful to know how long a dead body has been resting at a location.

Method

The soil microbial flora might be assessed by either molecular profiling techniques as detailed above or by culturing them using traditional methods. A dead animal or tissues at varying stages of decay would be placed on the surface of the ground and the underlying soil sampled at set intervals. Changes in the microbial flora could be estimated at varying levels of detail from the total number of culturable bacteria, fungi and protozoa by studying how the molecular profile alters with time. Any changes could also be linked to other soil physical and chemical characteristics, such as pH, organic matter content, nitrogen, phosphate and protein levels, as well as different soil types (e.g. comparing a heavy clay soil with a sandy loam).

5 Protists, fungi and plants in forensic science

Chapter outline

Protists as forensic indicators
Fungi as forensic indicators
Plants as forensic indicators: wood; pollen and spores; fruit, seeds and leaves
Plant secondary metabolites as sources of drugs and poisons
Illegal trade in protected plant species

Objectives

Explain how algae, and in particular diatoms, can provide an indication of drowning and also associate a person with a locality.

Discuss the potential and limitations of fungi as forensic indicators.

Describe how the characteristics of woody plants can provide forensic evidence in cases ranging from forgery to murder.

Review the potential of pollen analysis in linking a person or an object to a locality at a particular time of year.

Describe the variety of ways in which fruit, seeds and leaves can provide forensic evidence.

Discuss the criminal use of secondary plant metabolites and how these may be detected.

Discuss how forensic evidence of the illegal trade in protected plant species may be obtained.

Introduction

Plants form part of our diet and the evidence of what we have eaten and when we ate it is contained within our digestive system and our faeces. In addition,

Essential Forensic Biology, Alan Gunn
© 2006 John Wiley & Sons, Ltd

we are always coming into physical contact with plants in our homes, gardens and the wider environment; the evidence of this can be seen in the damage caused to the plants and/or bits of them that get attached to our hair, clothing or possessions. Plants are therefore potentially useful sources of forensic evidence for associating a person, vehicle or object with a locality. Despite this, the forensic potential of plant-based evidence is poorly exploited and there is much work to be done in this area. Protists and fungi are also extremely common in the environment but their use in forensic science has received even less attention than that of plants (see Table 5.1).

Protists as forensic indicators

The kingdom Protista is an artificial grouping containing a hugely diverse assortment of eukaryotic organisms, i.e. they all possess a membrane-bound nucleus and membranous organelles. Many of the Protista are not even closely phylogenetically related to one another. The amoebae belong to the phylum Rhizopoda, which contains about 200 species, of which the parasitic species *Entamoeba histolytica* is the most well known as the cause of amoebic dysentery. Most amoeba species are, however, free-living and they are common organisms in both terrestrial and aquatic ecosystems. Testate amoebae (Figure 5.1) possess cell walls that are coated with a protective covering – the 'test' – that is either secreted by the amoeba or composed of material gathered from the environment. The shape and size of these tests vary between species. They are particularly common in mires and peaty soils. Although there are few published

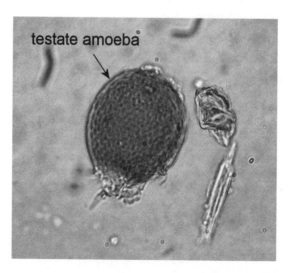

Figure 5.1 Light microscope photograph of a terrestrial testate amoeba

Table 5.1 Summary of forensic information gained from protists, fungi and higher plants

Sample	Forensic evidence
Testate amoebae	Geographical location Habitat
Diatoms	Geographical location Habitat Time of year Cause of death (e.g. drowning)
Fungi / fungal spores	Geographical location Habitat Time of year Poisoning Illegal trade in drugs
Whole plant	Damaged plants can indicate the site of a struggle or tread marks from shoes or vehicles Amount of healing can indicate time since damage occurred Time since burial (e.g. growth on grave site) Illegal trade in drugs (e.g. marijuana) Illegal collection and trade of protected species
Plant leaves	Geographical location Habitat Diet / last meal (e.g. fragments in gut contents or faeces) Illegal trade in drugs Illegal collection and trade of protected species
Plant roots	Time since burial (damage, repair and subsequent growth through or around skeleton or object)
Wood	Tool marks of object causing damage Amount of healing can indicate time since damage occurred Cause of death (splinters left in wound may identify murder weapon) Illegal trade in protected species Fraud (misrepresentation: object made out to be older than it really is or made from inferior wood)
Seeds	Geographical location Habitat Time of year Diet / last meal (in gut / faeces) Poisoning Illegal trade in drugs Illegal trade in protected species Fraud (e.g. mislabelling, contains GM material)
Pollen	Geographical location Habitat Time of year Poisoning (some flowers are poisonous) Fraud (e.g. mislabelling, contains GM material or from a different country of origin)

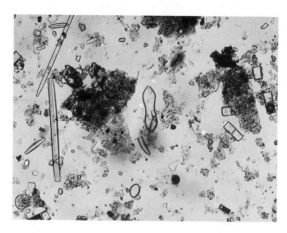

Figure 5.2   Mixed species of freshwater diatoms

reports of testate amoebae being used in forensic studies, they are common soil microorganisms and, like diatoms and pollen, their abundance and diversity could be used to characterize soil samples and thereby link material discovered on a suspect with a crime scene.

The term 'algae' encompasses a wide variety of organisms ranging from single-celled protests, to colonial forms, to huge multicellular seaweeds such as kelp. Most, but not all, algae contain chloroplasts. Although algae are ubiquitous in terrestrial and aquatic ecosystems, only the diatoms tend to be used to any great extent in forensic analysis. Diatoms are unicellular algae belonging to the phylum Stramenopila (phylum Bacillariophyta in some textbooks) and are characterized by the possession of beautifully patterned cell walls made out of silica called 'frustules' (Stoermer and Smol, 2001). They can be found in fresh water and salt water, as well as the surface of moist terrestrial habitats (Figures 5.2 and 5.3). Because they use silica extracted from their environment to build their frustules, their growth is limited by the availability of dissolved silica in the water. Each frustule is composed of two halves, which are referred to as valves, one of which is slightly bigger than the other so that the two fit together one inside the other. Each species of diatom has its own unique frustule design – a feature that facilitates their identification. Although most frustules dissolve when the algae die, their silicaceous nature means that some are preserved, especially if the conditions are favourable. Indeed, some soils, called diatomaceous earths, are composed almost entirely of fossil diatoms. There are numerous species of diatoms: more than 5600 have been described and there are probably more than 100 000 in total, and their abundance and species composition vary between locations owing to different diatoms preferring different conditions such as temperature, salinity and pH and the consequences of inter-species competition. Therefore, the presence of individual species or the species composition in a forensic sample can provide an identifying feature of a habitat or

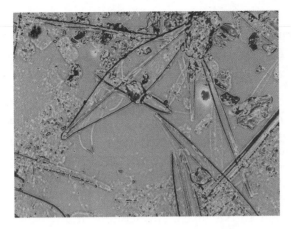

Figure 5.3 Freshwater diatoms. Note the large differences in size

location at a particular time of year. The rate of colonization of submerged corpses by diatoms and other algae has been investigated as a means of estimating length of submersion but there are few studies in this area (Casamatta and Verb, 2000).

Drowning results from suffocation following the immersion of the mouth and nostrils in a liquid, although some pathologists would state that it also involves the aspiration of fluid into the air passages of the lungs. The mechanism of death is very complex and not simply a consequence of asphyxiation. The diagnosis of drowning at autopsy is notoriously difficult and signs of immersion merely indicate that a body has been underwater (or some other fluid) for a period of time and not that the person died of drowning. Because diatoms can be found in many water sources, their presence in the lungs and other tissues has been used as an indication that the victim may have drowned (Pollanen, 1998). Diatoms can be recovered from the stomach (indicating that water was swallowed), from the lungs (indicating that it was aspirated) and also from the blood, major organs and the bone marrow (indicating the water was aspirated whilst the victim was still alive). Owing to their small size and the damage caused to the lungs during drowning, diatoms may pass through the alveoli and be swept around the body in the blood stream (Lunetta *et al.*, 1998). However, once the blood circulation ceases, any diatoms entering the lungs with water would not be transported elsewhere. By comparing the abundance and species profile of the diatoms found in the body with that of diatoms found in the river – or wherever the victim was recovered – it is possible to provide corroborating evidence with results from the autopsy to determine whether or not he/she drowned and, if so, whether it was at that location (Ludes *et al.*, 1999). For example, the presence of freshwater diatoms in a body recovered at sea would suggest that the victim may have died in a river and subsequently been swept out. The possibility that a person may have died in a water body other than the one in which they were found needs to be borne in mind when deciding on extraction techniques. For example, the frustules of marine diatoms are dissolved by Soluene-350 (a solubilizing agent), whereas the frustules of freshwater diatom species are not (Sidari *et al.*, 1999). Like all biological evidence, the results of diatom analysis need to be considered in context. For example, some workers question the sensitivity of the diatom test and the absence of diatoms does not mean that drowning did not occur, nor does the presence of diatoms in the body tissues mean that it did. Although diatoms are extremely common, they are not found in all water sources and even if they were present they may not find their way into the body organs or be recovered if they did. Similarly, the abundance of diatoms means that contamination at the time the body was retrieved, during the autopsy or during laboratory analysis is always a possibility unless extreme care is taken. Furthermore, if a person repeatedly swims in a lake or the sea, it is possible that they may accumulate diatoms within their tissues over time, so finding them does not mean that the person drowned there (Taylor, 1994). Diatoms may also be recovered from the tissues of persons who do not die of drowning or swim regularly. Diatoms are found in numerous man-made products ranging from building materials to the powder used in rubber gloves and can therefore be breathed in, and there is also a possibility that they might be absorbed through the gastrointestinal tract when consumed with foodstuffs – although the extent to which this occurs needs to be confirmed by further research. Consequently, diatoms can be found in the tissues of persons who died of causes other than drowning. However, it is generally accepted that finding diatoms within the bone marrow provides good corroborative evidence of drowning and identification to species level can exclude those that are contaminants (Pollanen, 1997; Pollanen *et al.*, 1997).

Murderers seldom succeed in killing their victims by drowning unless there is a big discrepancy in their size and strength or the victim is incapacitated by drink, drugs or disease and unable to resist. It also occurs after a victim has been assaulted and rendered unconscious or weakened and thrown into a body of water. It can be a cause of suicide and also infanticide. For example, in May 2002, a young Californian mother phoned the local police claiming that her 15-month-old child had been snatched from her by a stranger whilst she had been standing by a park fountain. The child was found later floating on the surface of a local river. The absence of physical trauma on the body made the police suspicious and an analysis of the water found in the child's stomach demonstrated the presence of diatoms characteristic of both those found in the fountain and the river. The police therefore suspected that she may have attempted to drown the child herself in the fountain and then thrown the body dead, or nearly so, into the river and concocted a story to cover her crime. Presented with this evidence, the mother confessed to having killed the child herself. This case demonstrates that evidence need not be overwhelming to encourage a confession. The presence of large amounts of water in the stomach is a good indication that a person was alive at the time of immersion but it is by no means conclusive proof of drowning. A toddler playing in shallow water often ends up falling over or attempting to drink the water regardless of the parents' efforts.

Haefner *et al.* (2004) have described how algal growth can be used to determine how long a dead body (in their case, that of a pig) or an inanimate object, such as a tile, has been submerged in water. Rather than attempt to identify the species of algae present and their relative abundance, they determined total algal density by measuring the amount of chlorophyll *a* present in a sample collected in a custom-designed sampling device (see Chapter 9). So-called 'blue-green algae' are not algae at all – they are prokaryotic organisms (i.e. they lack an enclosed nucleus and membrane-bound organelles) and are more correctly known as cyanobacteria. Despite being extremely common in both terrestrial and aquatic ecosystems their forensic potential does not appear to have been investigated, although signs of colonization of exposed bones is said to become apparent to the naked eye after 2–3 weeks under suitable conditions (Haglund *et al.*, 1988).

Fungi as forensic indicators

Mushrooms, toadstools and other fungi are commonly thought of as plants but in fact they belong to a totally separate kingdom of their own – kingdom Fungi – and their study is a separate science called mycology. Although many species of fungi are important in decomposition processes, they are seldom used in forensic studies of human and animal remains. Carter and Tibbett (2003) have attempted to remedy this situation by considering a wide range of fungi and their potential as forensic indicators, although Bunyard (2005) has commented that some of their suggestions are a bit too enthusiastic. In particular, many of the fungi they mention are small and difficult to identify. Although *Hebeloma syriense* has earned itself the moniker 'the corpse finder', there are few published scientific data to confirm this as most fungi found associated with corpses are also found in other substrates containing high ammonia levels. Similarly, the presence of a body close to the soil surface will be more readily identified by its smell, the disturbed soil and the frantic activity of invertebrate detritivores than by detection of fungi. There appear to be no published studies on how fungal populations change as a body decomposes and their effectiveness in determining the minimum time since death probably awaits the discovery of more rapid and simpler fungal identification techniques.

Fungi produce spores as a means of dispersal and this may occur as part of either sexual or asexual reproduction, depending on the species. Fungal spores are often produced in large numbers and, like pollen (see later), can have characteristic morphologies that facilitate their identification. Because some fungi have restricted distributions, specific ecological requirements and/or produce spores only at certain time(s) of year, their presence may be useful in associating a person or object with a locality.

A number of mushroom species are highly poisonous and may cause death if treatment is not obtained soon enough. It is therefore to be expected that they have been exploited by murderers from time to time. The Roman Emperor Claudius was reportedly the victim of intentional mushroom poisoning but there are relatively few accounts of their being used to murder someone since then. In the absence of witnesses or other evidence, poisoning by any means can be difficult to detect because the forensic pathologist may have no reason to suspect that it was the cause of death and would therefore not send tissue samples for analysis. In continental Europe, where the picking of wild mushrooms is popular, poisonings are common as a result of mistaking poisonous for edible species. For example, in 2000 there were 2740 reported cases of mushroom poisoning in Russia. In the UK, mushroom poisoning is more likely a consequence of people confusing wild-picked magic mushrooms, of which there are around 12 different varieties, with poisonous ones. The liberty cap mushroom (*Psilocybe semilanceata*) is the most commonly consumed 'magic mushroom' and the laws surrounding its use are decidedly odd. Growing, gathering or possessing the fungi does not contravene any laws, however, any attempt at preparation (such as cutting, drying, powdering or freezing and packaging) renders it a Class A controlled drug. In fact the situation is even more bizarre because operatives openly selling magic mushrooms in the UK (some of it imported from abroad under the supervision of Customs and Excise) have recently been told that their mushrooms are subject to backdated 17.5 per cent VAT because they are sold for their effect as a drug rather than as a food (*The Independent*, 10 August 2004).

There are numerous species of moulds and most of them live a perfectly blameless existence outdoors where they are involved in decomposition processes. A few species, however, are also found in human habitations and some of these can present a serious health hazard, cause food to be condemned or damage wooden structures and household furnishings. Although measures can be taken to prevent or limit the growth of moulds, where these fail or are not put into practice the consequences can be serious. Identifying and linking a species of mould with a person's health problem, food condemnation or household damage could be important in a legal case (Money, 2004).

Plants as forensic indicators

The kingdom Plantae includes all multicellular terrestrial organisms that carry out photosynthesis and have an embryo that is protected within the mother plant. It therefore includes organisms as diverse in size and shape as mosses, potato plants and oak trees. For the purposes of this chapter, it is not intended to take a strictly taxonomic approach to the role of plants in forensic science.

For example, there is no phylogenetic grouping consisting solely of trees but it is convenient to deal with them under a single heading.

Wood

The study of wood can provide information applicable to many aspects of forensic science, ranging from the trade in protected species and the fraudulent sale of copies of valuable antiques to identifying the weapon(s) used in a murder. Some species of tree produce wood with a distinctive texture and smell. For example, in October 2004, 33 tons of wood from the Brazilian tree *Dalbergia nigra* (Palosanto de Rio) were impounded in Spain where it had been imported illegally for the manufacture of high-quality Palosanto guitars. Police became suspicious of the increasing supply of Palosanto guitars coming onto the market despite trade in the wood being highly controlled since 1992. Following a surveillance operation, the wood, identifiable from its characteristic properties, was found hidden in batches of legally imported hardwood timber. Colour, texture and smell are, however, not an invariably accurate means of identifying wood so microscopic analytical techniques are more normally employed (Figure 5.4). On their own, even these techniques are seldom sufficient to enable species identification, although it may be possible to determine the genus or sub-generic grouping. A more accurate identification may be possible if more information, such as country of origin, is known. For example, Scots Pine, American red pine and several Asian red pine species cannot be distinguished from one another on the basis of microscopic anatomy but if one knew that the wood came from a tree growing in North America then the species would almost certainly be American red pine because it is the only red pine species native to the region. The feasibility of using molecular techniques for species identification from wood samples is being investigated but they are not yet in widespread use. Wood fragments can prove useful in associating an object with a crime. For example, in an Australian case, wood fragments found embedded in the skull of a murder victim were identified as coming from a species of rubber tree grown in southeast Asia. The wood from this tree is used in the production of baseball bats so the police then knew what sort of murder weapon they were looking for. Furthermore, because only certain makes of baseball bat are prepared from this type of wood, the search was narrowed down further and ultimately they were able to identify the actual bat that was used as the murder weapon and connect it to the crime.

If one looks at the cross-section of a typical tree trunk, one can see a pattern of concentric dark and light rings. These rings result from the distinction between early wood produced in the spring and late wood produced towards the end of summer or early autumn. The early wood is characterized by large cells with thin cell walls, whereas the late wood has small cells with thicker cell walls (Figure 5.5). The dividing line between the late wood of one year and the

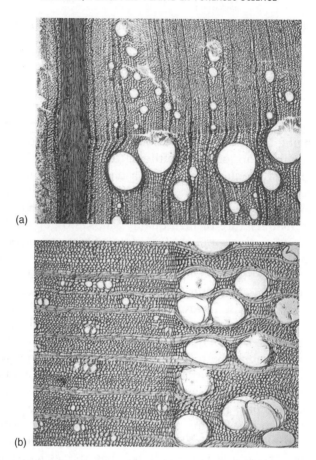

Figure 5.4 Although wood from different trees often looks very similar, its microscopic anatomy may be quite distinctive. Compare these samples of (a) European Oak (*Quercus robur*) and (b) Ash (*Fraxinus excelsior*). For microscopic analysis, the samples should be cut across and along the grain of the wood and then compared to reference material

early wood of the succeeding year is seen as the line between growth rings. In temperate climates such as the UK, trees produce only one growth ring per year and a tree's age can therefore be determined from the number of rings. In addition, the width of the rings provides an indication of the environmental conditions at the time, with thick rings reflecting good growth conditions. The study of tree growth rings is known as dendrochronology and it has proved extremely useful in determining climate change over the years. In forensic science, the study of tree rings can provide evidence of the age of an object. Typically, the age of a piece of wood can be estimated by comparing its ring patterns with those of samples obtained from the geographical region it is claimed to have originated from. For example, if a painting is made directly onto a piece of wood it is possible to determine the date when the tree it came from was felled, provided that

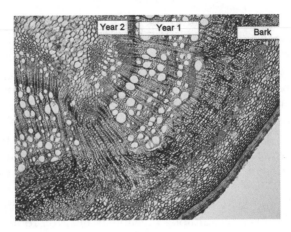

Figure 5.5 Cross-section through a 2-year-old oak (*Quercus robur*) sapling showing the formation of growth rings

sufficient rings are available for analysis. Consequently, it is also possible to state whether the artist to whom the work was being attributed would have been alive and at an appropriate age to have executed the work. For example, a painting attributed to a medieval artist produced on a piece of wood that came from a tree felled in 1865 could be nothing other than a fake. This is clear cut when the object in question is made out of a single piece of wood but in the case of antique furniture and musical instruments things may be more complicated owing to repairs and alterations that have been undertaken over the years. For example, all Stradivarius violins, without exception, have had their original necks sawn off and modifications made as a consequence of changes in playing techniques. Furthermore, although dendrochronology, like most of the other forensic techniques used to determine the provenance of paintings and antiques, is good at proving falsification it cannot necessarily prove authenticity. It should also be borne in mind that modern fraudsters are often aware of the forensic procedures used to detect their work and will use techniques such as making picture frames out of old wood and ageing nails in salt solution. When examining furniture and similar items, it is the less visible parts (such as the back or the corner blocks) that can provide most information on the country of origin. Because these parts are not seen they tend to be made from cheap local wood, whereas the exterior ornate parts are often made from imported timber.

The study of tree rings may also be useful when a tree or sapling is found to have grown through a skeletonized body or within a grave: the number of rings indicates how long the body has rested at the site. Courtin and Fairgrieve (2004) have described an interesting case in which the growth of a tree branch was affected by the presence of a dead body lying across it. The pressure of the body resulted in an asymmetrical pattern of growth rings and computerized quantification of these rings enabled a back-calculation to when the body came to lie across the branch. Tree growth rings take time to form, so the technique could only state that the body came to rest there at some point within a 10-month period but it subsequently transpired that this encompassed the time during which the dead person disappeared. The technique could be useful in cases in which persons are left bound or hanging from a tree and more accurate evidence of the postmortem interval (e.g. physical changes, insect activity, etc.) was lacking. Roots also show annual growth rings and therefore, where these have penetrated the clothing or bones, they will indicate how long the body has been present at the site, as will the depth to which roots have penetrated the body (Willey and Heilman, 1987). Plants can cause a variety of postmortem artefacts and these should be distinguished from damage caused at the time of death. For example, although seemingly delicate, roots and shoots can exert enormous pressures that can be sufficient to force apart even the sutures of the skull. Similarly, roots growing across the surface of or through bone can cause the formation of a characteristic network of intricate and ramifying grooves as a consequence of their secretions (Haglund *et al.*, 1988). The presence of plant tissues and their relationship to human remains should therefore always be taken into account.

When a gun is fired, its bullets acquire a characteristic pattern of markings from the bevelling. In a similar manner, every hammer, saw, chisel or axe has its own unique surface properties and, when used, these are faithfully preserved in substrates such as wood or bone as 'tool marks'. Any hard material causes such marks when it hits or is dragged across a softer material and their study is known as 'tool mark analysis'. Flesh wounds do not retain detailed impressions of the object that caused the injury, although it may be possible to determine the shape, length and width of the weapon used. If the underlying bone is damaged, a better record may be preserved. Tool marks take the form of a negative impression of the object (i.e. 'the tool') that inflicted the damage, from which can be deduced its shape and size from the contours and rows of parallel grooves caused by the object being dragged across the surface. It is therefore possible to link a suspect with a crime by comparing the cutting surface of his tools with marks found at the scene. For example, in May 2003 part of the evidence that was used to convict a St Helens man of illegally collecting the eggs of wild birds was the discovery in his possession of a chisel that produced marks identical to those found on a tree in a local park where the nest of a great spotted woodpecker had been broken into and the eggs removed.

Perhaps the most famous case in which wood analysis was central to the prosecution was that of Bruno Hauptman, who was arrested for the kidnap and murder of the son of the American aviator Charles Linbergh. The crime was committed with the aid of a crude homemade wooden ladder that was left behind at the time of the kidnap. By comparing the milling marks left on the various components of the ladder it was possible to trace each of them back to their original supplier. It was also noted that one of the ladder rails was made from low-grade unweathered wood, such as that used inside a house or building, and it had four distinctive square-shaped nail holes. Bruno Hauptman was initially arrested when he used money from the kidnap ransom; when police searched his property they discovered that one of the floorboards in the attic was eight foot shorter than the others and that the four square nail holes found in the ladder rail mentioned above were a match for the holes in one of the attic floor joists. Furthermore, the annual growth rings of the wood used in this ladder rail matched those of the short floorboard. Within Bruno Hauptman's workshop the police also found a hand plane with a dull damaged blade that produced marks identical to those found on the ladder. This, together with other evidence, led to Bruno Hauptman being found guilty and he was executed in 1936.

Pollen and spores

Primitive plants such as mosses (bryophytes) and ferns (pteridophytes) do not produce seeds but they do form spores (Figure 5.6). These spores are very small and are shed in vast numbers to be dispersed by air currents. If a spore lands in a suitable spot, it germinates and produces a gametophyte – the structure in a plant's life cycle that produces gametes. The spores of the club moss, *Lycopodium clavatum* are commonly used as a lubricant during the manufacture of condoms to prevent them sticking to themselves when they are rolled up; other plant products such as potato starch and corn starch are also used for the same purpose. It has therefore been suggested that the presence of these substances in vaginal swabs can be used as forensic indicators in cases of sexual assault in which the man has used a condom (Berkefeld, 1993; Blackledge, 2005); it is reported that lycopodium spores can be detected in vaginal swabs

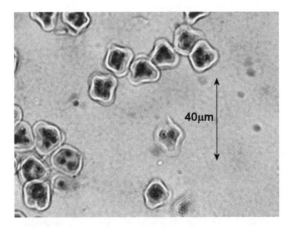

40µm

Figure 5.6 Spores of the moss *Polytrichum* sp.

up to 4 days after sex has taken place (Keil *et al.*, 2003). This information could prove useful in a number of situations. Firstly, although not a foolproof method of avoiding leaving traces of semen, the wearing of a condom reduces the likelihood considerably. Consequently, in cases in which there is a dispute whether a sexual act took place but there is no DNA evidence, the presence of lycopodium spores, etc. in a vaginal swab would indicate that the woman had had sex in the recent past with a man wearing a condom and it would there-fore be worthwhile pursuing the investigation. This would also be the case if a dead woman was found and sexual assault was suspected but semen could not be detected and medical evidence was inconclusive. In addition to the presence of spores and other particulate matter, condoms also leave a 'chemical signa-ture' formed from traces of chemicals used in their manufacture (lubricants, spermicides, etc.) that identify the brand. Men are creatures of habit and a condom-using serial sex offender is therefore likely to use the same brand in most if not all of his crimes. Therefore, identifying his brand to choice can be useful in building up a profile and in the search for associative evidence.

Pollen is the term used to describe a collection of pollen grains and these are the male gametophytes formed by seed-producing plants (i.e gymnosperms such as pine trees and other conifers, and angiosperms such as grasses, rushes, but-tercups, daisies and lime trees). In order for fertilization to occur, the pollen grains have to be transferred from the 'male' part of the plant, where they are produced, to the 'female' part of the same, or different, plant in a process called pollination. Pollination is affected by a wide variety of agencies, including the wind, insects, birds and mammals, and the morphology of pollen grains has evolved appropriately. For example, plants that are wind pollinated produce pollen grains that have a smooth surface (Figure 5.7) although a few, such as those pine trees, have lateral air sacs to aid their dispersal (Figure 5.8). They are released in enormous numbers owing to the low chances of any individual grain reaching its target. Plants pollinated by insects or other animals tend to produce fewer pollen grains and sometimes these are covered with projections or sticky secretions that facilitate their attachment to the pollinator (Figures 5.9 and 5.10). The outer layer of pollen grains is called the 'exine' and is extremely resistant to degradation. This, together with the characteristic morphology of some species of pollen, means that it can be identified thousands of years after it was originally produced – a feature commonly used in archaeology and palaeobiology when reconstructing the vegetation of past landscapes. The study of pollen grains and spores is known as palynology, although researchers in this area commonly also study fungal spores and microscopic (5–500 μm) plant and animal structures collectively known as palynomorphs. Palynomorphs are living or fossil structures resistant to the strong acids, etc. used to extract pollen. They include organisms such as testate amoebae, dinoflagellates and diatoms.

Because many types of pollen and spores can be readily identified, are pro-duced at specific times of year and they are usually well preserved, they have forensic potential as a means of associating people or objects with a locality.

(a)

(b)

Figure 5.7 Flowering head of the grass cocksfoot (*Dactylis glomerata*) (a) and its pollen (b)

The pollen of wind-pollinated plants and spores of some ferns (e.g. bracken, *Pteridium aquilinum*) are formed in such huge numbers that any object within the general area will be contaminated with what is sometimes called the 'pollen rain'. Pollen from wind pollinated plants is also breathed in and can be recovered from the lungs and is also swallowed and found in the gastrointestinal tract. Cannabis plants (*Cannabis sativa*) are wind pollinated and if grown indoors their pollen will be found on every surface in the area. Consequently, evidence of past presence remains long after the plants are removed. Even if the grow room is thoroughly cleaned, traces of pollen can be recovered from cracks and crevices, air extractors, the tops of doors, etc. Similarly, the clothes of anyone working with cannabis plants will be contaminated with the pollen. Cannabis products derived from plants grown outside will contain both cannabis pollen and that of other species of plants found in the area. This can provide an indication of the region or country of origin. By contrast, products

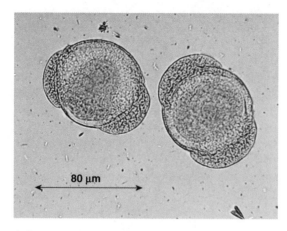

Figure 5.8   Pollen of the Scots Pine tree (*Pinus sylvestris*) – a wind-pollinated plant. Note the lateral air sacs

derived from plants grown indoors in a well-sealed grow room will be free of pollen from other plants (Stanley, 1992).

In September 2001, the headless, limbless torso of a 4–6-year-old boy was recovered from the bank of the river Thames, sparking one of the most detailed forensic investigations in recent years and one in which plant evidence has proved of crucial importance in guiding the investigation. In the absence of any identifying features, detectives christened the boy 'Adam' – which helped those working on the case and kept the victim's international media profile high. A combination of stable isotope analysis and DNA studies indicated that the boy probably originated from a region of Nigeria. The next fact to be determined was whether the boy was dead before he reached the UK and, if not, how long he remained alive once he got here. The child's stomach and upper intestines were empty, indicating that he had not eaten for some days before he was killed. However, it was possible to extract pollen grains from his lower intestine and these included those of plants typical of the UK and North West Europe, such as alder, suggesting that he had been alive in the UK and therefore breathing the pollen in before he was killed. Similar pollen could not be extracted from supermarket foods, indicating that the pollen was breathed in rather than con-sumed in the food. It takes about 72 hours for food to pass completely through the gut, so the presence of food (and pollen) in the lower intestine suggested that the boy had been alive here for about 48–72 hours. Also present in the lower intestine were fragments of the outer seed layers of the calabar bean (*Physostigma venenosum*). The beans of this plant are extremely toxic because they contain physostigmine (eserine), which is a potent inhibitor of cholinergic nervous transmission; a single ripe bean is capable of killing a man. It is thought that the bean(s) may have been administered to the boy to sedate him before he was killed. Calabar beans used to be employed in West African rituals in which a person suspected of a crime was forced to eat the bean: if the person

(a)

(b)

Figure 5.9 Flowers of yellow rattle (*Rhinanthus minor*) (a) and its pollen (b). Bumblebees normally pollinate this plant although it can self-pollinate. Note that some insect-pollinated plants produce relatively smooth pollen grains

died he was considered guilty and if he vomited the bean he was innocent. The case has not yet been solved and one line of investigation is the possibility that Adam was the victim of a 'muti killing' in which humans are sacrificed to propitiate spirits and to bring good luck.

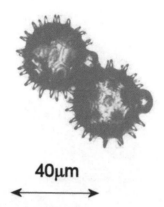

40μm

←——————————→

Figure 5.10   Pollen of coltsfoot (*Tussilago farfara*) – an insect-pollinated plant. Note the projections to aid it sticking to the body of the pollinating insect

When released from plants, pollen varies in its dispersal capabilities and this can be measured as its rate of fall or 'sinking speed'. This is influenced by a combination of biological factors (such as the size and shape of the pollen), environmental factors (such as the weather) and local topographical features (such the presence of hedges or valleys). Maize (*Zea mays*) pollen is relatively large (90–125 μm) compared to other wind-pollinated grasses (usually ~40 μm) and has a fast sinking speed so the presence of large amounts in or on an item would suggest that it had spent some time situated close to a flowering maize crop. By contrast, birch (*Betula* spp.) pollen is much smaller (20–30 μm) with a slow sinking speed and therefore its pollen gets distributed more widely. Many plants release their pollen over restricted time periods, so its presence can act as an indicator of the time of exposure. For example, maize pollen is released over a period of 2–14 days, although 5–8 days is more usual. Wind-pollinated plants also tend to release most of their pollen in the early morning when upward air currents are most pronounced. The presence of pollen from animal-pollinated plants on a person or clothing usually requires direct contact to be made, e.g. a person may handle a bouquet of flowers (lillies are notorious for staining clothing with their pollen) or lie on the ground or walk through a field where flowers are growing.

Pollen may become associated with food, the most obvious example being honey (Figure 5.11). There is an active trade in the sale of mislabelled honey in the European Union, particularly since honey from China was banned in 2002 owing to concerns about the levels of antibiotics such as chloramphenicol that were being used by beekeepers to prevent diseases among their bees. Following the ban, it is alleged that Chinese honey was smuggled into India where it was repackaged and sold on. Suspicions were also raised about how Singapore somehow increased its honey production from virtually zero to become the world's fourth largest honey exporter at exactly the same time that Chinese

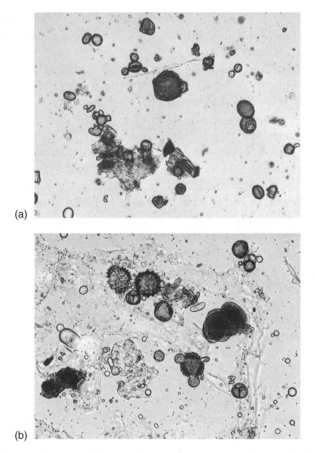

(a)

(b)

Figure 5.11 Pollen analysis can indicate whether a honey's source is correctly labelled. Compare the pollen in (a) Armenian wild flower honey and (b) Greek 'wild mountain honey from herbs and coniferous trees'

honey was banned. The problem of honey containing antibiotics is by no means confined to China and they have been found in honeys from many other countries as well. The EU ban on the import of Chinese honey has been rescinded but there are still concerns about mislabelling. The presence of pollen in honey can help identify its country of origin and, in the case of single flower honey, whether this is a true reflection of its composition. For example, in July 2004 a Northumbrian beekeeper was successfully prosecuted for marketing honey as being produced in the Scottish Borders when it actually originated in Argentina. There are therefore concerns about the production of ultra-filtered honey because although this treatment would remove contaminants such as antibiotics it would also remove the pollen that facilitates the honey's identification. Because honey is marketed as a 'natural' wholesome food there are concerns about it becoming accidentally contaminated with pollen from genetically modified (GM) crops and this will no doubt become a source of legal dispute as GM crops become widely grown.

A great deal of pollen ultimately comes to rest in the soil and because plants differ in the amount of pollen they produce and its distribution characteristics a soil's pollen profile is an identifying characteristic. This can be used to associate soil samples found on shoes, clothing or spades and other digging implements with the scene of a crime (Horrocks *et al.*, 1998, 1999). A classic case in which this approach has been used is the trial of Bosnian war criminals for the genocide perpetrated at Srebrenica in the summer of 1995. When the town of Srebrenica fell to the Bosnian Serb army Drina corps, which was under the command of General Radislav Krstic, about 7000 men were led away and executed and their bodies buried in four or five mass graves. Some of the victims may have been buried alive. After about 3 months, the killers attempted to dispose of the evidence by digging up the graves and dispersing the bodies among numerous 'secondary burial sites' many miles away. In this way it was hoped that the deaths would appear to be a consequence of small local skirmishes rather than a carefully planned massacre. However, several years later, forensic scientists compared the pollen profiles from soils at the primary burial sites, the secondary burial sites and the soil within the skeletal cavities and surrounding the remains at the secondary sites and they were able to prove conclusively that the bodies had been moved. For example, the presence of large amounts of wheat pollen associated with the soils within and surrounding one batch of victims found in their secondary grave did not match that of the soil in the immediate vicinity but did match that of the primary execution and grave site. Similarly, the pollen profile of another batch of victims indicated that initially they been killed and buried close to a broadleaved woodland containing oak, birch and hornbeam. Other cases in which pollen analysis has been used in forensic analysis can be found at http://www.dal.ca/~dp/webliteracy/projects/forensic/palynology.html and http://www.crimeandclues.com/pollen.htm.

Pollen analysis does, however, have its problems. It can be difficult to identify some types of pollen, especially that of grasses, to species level. Therefore, there can be problems with taxonomic resolution. Furthermore, some types of pollen degrade faster than others, e.g. oak pollen degrades faster than lime. Consequently, the pollen profile may not accurately represent the plant species composition of the site. In general pollen profiling is most effective for identifying a location if the pollen includes examples of rare plants or unusual combinations of plant species.

Fruit, seeds and leaves

The fruit, seeds and leaves of many plants form part of our diet. Humans are unable to digest the cellulose that makes up the walls of plant cells or the tough outer coating of many seeds, so these structures pass through the gut and may be identified from their characteristic morphology. Consequently an analysis of a person's stomach contents and/or faeces can provide evidence of what they have been eating and therefore, potentially, where they ate it. For example, in an American case in which a young woman was murdered, some witnesses claimed that her last meal was a lunch of burger with onions and pickles that she shared with her boyfriend at a McDonald's fast-food restaurant. However, when her body was found, her stomach contents included a much larger proportion of vegetable matter than this would have provided. The police therefore interviewed staff at a local Wendy's salad bar where a waitress remembered seeing the woman leave with another man after dinner. The state of degradation of the digestible components and their position within the gastrointestinal tract can provide a crude indication of how long it was since the person ate that meal but the rate of stomach emptying and passage of digesta through the gastrointestinal tract is affected by numerous factors. For example, the presence of large amounts of lipid and protein (such as that present in the traditional English breakfast of bacon and eggs) will slow gastric emptying, as can underlying medical conditions such as diabetes. Furthermore, there can be considerable variation between normal healthy subjects in the rate of transfer of food between regions of the stomach. Plant matter may be found in the body outside the gut. For example, aspiration (breathing in) of the gut contents during vomiting is a fairly common occurrence in persons suffering acute alcohol intoxication and drug overdoses and may prove fatal. In these cases, vegetable matter may be observed in the airways of the lungs either macroscopically or in histological sections.

A number of plants, such as goosegrass (*Galium aparine*, Figure 5.12, also known as 'cleavers' or 'sticky willy'), wild carrot (*Daucus carota*) (Figure 5.13) and lesser burdock (*Arctium minus*) disperse by producing seeds that are armed with hooks designed to catch onto the coat of passing animals. They also latch onto clothing and hair and can be extremely difficult to remove. They therefore have forensic potential in associating a person with a locality and/or their activities, especially if a reliable DNA match can be made. An interesting case has been described in Germany (http://www.strate.net/e/publications/weimar.php3) in which the fully clothed body of a young girl was found in undergrowth: she had a regular distribution of goosegrass fruit over her whole body, including the inside of her trousers and on her underwear, amounting to 362 fruit in total. The number and distribution suggested that the girl did not pick up the fruit whilst playing – in which case they would have been concentrated on her legs. Goosegrass tends to grow among nettles so it is unlikely that any child would roll around on the ground where they occur. The presence of the fruit inside the trousers was considered suspicious because they are itchy and the normal reaction would be to stop whatever you are doing and remove them. This, in conjunction with other evidence, led the investigators to believe that the girl was not dressed in the trousers she was found in at the time of her death. Instead,

(a)

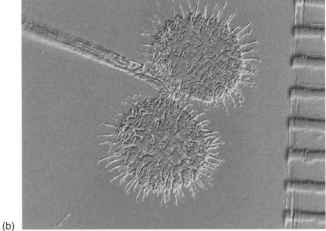

(b)

Figure 5.12 Goose grass (*Galium aparine*) growing among nettles (a). The mature seed heads (b) readily attach themselves to clothing whilst the spines (c) on the leaves and stem will capture fibres and hairs

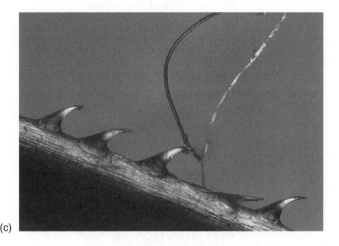

(c)

they think that the dead girl was transported to a site close to where she was found dressed only in her underwear. There she was laid down on the ground and dressed – a trampled area where goosegrass was growing was identified as the likely spot – and this would explain the even distribution of goosegrass fruit and the presence of the fruit inside the clothes.

Because seeds are larger than pollen, produced in smaller quantities and tend not to fall far from the parent plant, they can be helpful in connecting a person with a crime scene. However, like pollen, they are most useful if they include examples of rare plants or those associated with a limited distribution or particular habitat. It is also easier to extract DNA from seeds and this can help in the identification process.

Plants are often extremely easy to damage but very difficult to kill. Whenever a person or object comes into contact with a plant, there is a possibility that the plant will be damaged in some way and, if so, the evidence of this association will usually be preserved. For example, when a vehicle drives over grass, it leaves behind imprints from which one can determine the direction of travel and tyre characteristics. The latter, if preserved well enough, can be used to identify the vehicle responsible. Similarly, when walking across a newly mown lawn or field one usually picks up grass clippings on ones shoes and clothing and these have been used as forensic evidence (Horrocks and Walsh, 2001).

When a body or object is buried both above- and below-ground, parts of plants will be damaged. If the growth region, (known as the meristem) of a root is affected, a permanent scar will result and the number of growth rings formed after the scar will indicate how many years ago the damage was caused. Owing to the disturbance the burial site will be marked by a characteristic sequence of succession plants, beginning with those that specialize in colonizing bare soil. If a body has been buried in nutritionally poor soil, its decay will probably result

(a)

(b)

Figure 5.13 Dry seedhead of the wild carrot (*Daucus carota*) (a) and its seeds (b). Note the spiny projections to facilitate catching onto the hairs of passing animals

in the noticeably more luxuriant growth of plants upon and around the grave. When a plant, or part of one, is covered so that it is no longer able to photosynthesize, it starts to become yellow and etiolated and, presumably, it will be accompanied by changes in gene expression. This could provide an indication of how long a body or object has rested at a particular site, although there appear to be few forensic studies to quantify this.

Plant secondary metabolites as sources of drugs and poisons

Plants produce numerous chemicals that are not essential for basic growth and development, collectively known as secondary metabolites. They help protect the plant from diseases, have structural functions or are toxic and protect the leaves, seeds, etc. from being eaten by herbivores. Plants such as hemlock and foxgloves are notorious for their poisonous properties. Even food plants can be poisonous if grown or stored under sub-optimum conditions. For example, when potato tubers are exposed to light and turn green, they form large quantities of solanine (an alkaloidal glycoside), which is resistant to normal cooking and potentially lethal. However, green potatoes have a bitter taste and it is unlikely that a person would consume a large quantity. The use of plant toxins to intentionally poison someone is rare in the UK although accidental poisonings are more common – often the result of children consuming the berries of mistletoe, yew, etc. and the seeds of laburnum, which are seldom fatal. In some parts of the world plant poisoning is more common and in a study in the southwestern state of Kerala in India it was estimated that 10 per cent of all fatal poisonings were due to consumption of extracts prepared from the seeds of the tree *Cerbera odollam* (Gaillard *et al.*, 2004). The extract contains cerberin, a highly potent toxin that is capable of inducing a fatal heart attack. At autopsy, unless the pathogist has reason to suspect poisoning, the heart attack would be put down to natural causes. The plant extract has a bitter taste but this can be disguised with spices. Most of the victims are women and because the women often eat separately from the men in traditional households the serving of separate food is made easier. Women who marry into such households are sometimes subject to enormous pressures owing to dowry disputes, the demand of exacting standards of behaviour, the running of the household and the inability to produce a male heir. In extreme cases the woman may be murdered ('accidentally' setting themselves on fire while preparing food is remarkably common) or commit suicide. It now appears that poisoning – either homicidal or suicidal – may be far more widespread than previously thought and it is possible that they may be occurring in Asian communities living elsewhere in the world where pathologists would be even less likely to suspect that cerberin played a part in a person's death.

Considerable concern has been generated over the possible misuse of ricin as a biological warfare agent. Ricin is an extremely poisonous glycoprotein that can be isolated from the seeds of the castor bean plant *Ricinus communis*; properly prepared castor bean oil and pomace and the castor bean seed cake used as an animal feed are both harmless because they are processed to remove the toxins. The whole seeds are unlikely to pose a hazard because their coat is hard and impermeable, thereby preventing the toxin from diffusing out and being absorbed, even if the beans are swallowed. Similarly, there is little risk of the poison being absorbed through the skin when handling the seeds, although any dust formed through abrasion is potentially more dangerous if it is breathed in. Purified ricin poses much more serious risks than the castor bean seeds and could theoretically be used to contaminate food or water, or used as an aerosol. In 1978, the Bulgarian political exile Georgi Markov was assassinated in London by being injected with ricin. As he was walking along the street he was struck on the leg by someone with an umbrella, which was a common enough occurrence and he thought no more about it. However, in this case the umbrella had been modified to inject a small metal sphere and within the sphere was ricin. Four days after being injected Markov died of gastroenteritis and organ failure and although ricin was never detected in his body the coroner stated that this was almost certainly the poison that was used. Since this case there have been no recorded incidents of persons being intentionally poisoned with ricin, although the ease with which castor bean seeds can be obtained and the extreme toxicity of ricin provides it with a high media profile. In July 2004, the US Food and Drug Authorities reported that they had found two jars of baby food that had been tampered with and contained ground-up castor beans and trace amounts of ricin. Babies had already eaten some of the food before suspicions were aroused and it was tested, but they suffered no ill effects because the ricin had not been purified and was present in very low amounts.

Among the secondary plant metabolites are various chemicals that humans have used as drugs to achieve an altered mind state. Unfortunately, this is often accompanied by addiction and damage to the user's health. Drug abuse is a major cause of criminal behaviour so, not surprisingly, the detection of drugs and drug residues often forms part of a crime scene investigation. The analysis of suspect powders, body tissues / fluids, etc. for drug residues is a job for trained pharmaceutical chemists but because many of the drugs are derived from plants these specimens would fall under the remit of a botanist or biologist. Plant taxonomy is often looked upon (incorrectly) as a rather dull subject and has been quietly dropped from many undergraduate degrees in biological sciences over the years. However, it can generate intense, acrimonious debates and even apparently simple questions, such as 'what is a species', can prove extremely difficult to answer. The taxonomy of the genus Cannabis is one such problem area. Some botanists consider the genus to be monotypic, i.e. containing only a single species, whilst others state that it is polytypic, containing several species or subspecies. Such distinctions are important if a person is able to claim that he/she was not in possession of the illegal *Cannabis sativa* var. *indica* but another variety or subspecies. 'Skunk' is a variety of cannabis that has been developed for indoor growth and has sometimes been referred to as a quite different type of drug. However, there is little evidence that its potency is any different from normal herbal cannabis and it would be impossible to distinguish between the two using normal forensic techniques (http://www.idmu.co.uk/skunkfaq.htm). Potency varies between individual plants, how and where the plants were grown and how long and under what conditions the cannabis was stored.

The UK laws surrounding the possession and use of cannabis plants / products are complex. For example, provided that certain strict conditions are met, such as only small numbers of plants / amounts of drug being involved and the person admits guilt, there is no need for confirmatory forensic analysis or the use of a test kit such as the Duquenois-Levine colour test: the experience of a police officer trained to recognize cannabis from its texture, physical appearance and smell is considered sufficient evidence (http://www.homeoffice.gov.uk/docs/hoc9840.html). In more serious cases, e.g. the possession of large numbers of plants with intent to supply, where the defendant disputes the allegations or the investigating officer is in any way unsure about the identification, laboratory analysis is required. Drug testing kits will prove the presence or absence of a drug but it is often useful to obtain the more detailed information that can be supplied by gas chromatography linked to mass spectrometry (GC–MS). There is currently some interest in the use of DNA profiling to link samples of marijuana found on a suspect with the plants it came from. In this way it is theoretically possible to link the grower, the distributor and the user together. There are, however, potential problems because marijuana can be grown from cuttings and the resultant plants are therefore all genetically identical clones. Consequently, a match between an individual marijuana sample and an individual marijuana plant does not mean that either the plant or its grower was the source. The technique is not yet applicable to cannabis resin because it contains insufficient DNA.

Under UK law, cannabis cultivation may only be undertaken under license and is either for research / medical purposes or for the commercial production of hemp – which has an extremely low psychoactive content. Illegal production of cannabis can be punished with a maximum penalty of 14 years imprisonment, although the possession, sale and use of cannabis seeds for culinary purposes are not subject to the Misuse of Drugs Act 1971.

It is often necessary to determine the source of contraband drugs in order to prove importation (and therefore trafficking) and to identify the region supplying the contraband. A variety of techniques may be employed, including identification of insect and/or pollen contaminants and analysing the drugs for their stable isotope ratios and the presence of trace alkaloids. Cocaine samples can be related to their region of origin with over 90 per cent accuracy by analysing the ratios of ^{15}N to ^{13}C and the levels of trimethoxycocaine and truxilline (Ehleringer *et al.*, 2000). Regional differences in ^{15}N arise as a consequence of variations in the soil characteristics across the main coca growing regions of the South American Andean Ridge, whereas the ^{13}C levels vary owing to local differences in the length of the wet season and overall humidity. Stable isotope analysis techniques are also proving useful in cases of food fraud, such as when products are adulterated or their geographical origin is miss-stated because it is not commercially viable to artificially adjust the isotope ratios. However, to be effective, stable isotope techniques require extensive databases of authentic reference samples and certified reference materials to monitor the accuracy of the measurements.

Drugs are sometimes added to food either intentionally or unintentionally, which may have unforeseen consequences. For example, the consumption of poppy seeds used in the baking of poppy cakes, bagels, etc. can result in high levels of morphine and codeine in the urine that could be mistaken for drug abuse (Selavka, 1991). This has relevance owing to the increasing use of workplace drug testing, especially in the oil and transport industry. The only sure way of distinguishing between poppy seed consumption and heroin abuse is through blood tests because these are more specific. The addition of cannabis to cakes is a common practice among recreational drug users but is sometimes done with malicious intent and the cakes presented to unsuspecting recipients as a gift. In February 2004, ten schoolteachers working at a school in Lueneburg, northern Germany, were taken to hospital suffering from hallucinations, nausea, etc. after eating a chocolate cake spiked with cannabis, and similar cases have also been reported in the UK. Drug intoxication can lead to accidents, especially when driving, and the possibility that a person may have been the victim of 'spiked' food needs to be considered.

Illegal trade in protected plant species

The world market in the illegal trade in protected plant species is probably as widespread and lucrative as that in protected animals although it attracts far less publicity. Some plant species are now nearing extinction in the wild as a consequence of over-collection. For example, cycads are primitive plants, many of which thrive in hot, dry climates that have become hugely popular for their decorative properties – and sheer rarity. Prize specimens can trade hands at over £20 000 each and an illegal trade worth millions of pounds a year is thought to operate between South Africa and North America. As a result, of the world's 298 described species, over half have become endangered and some are now thought to be extinct. Specimens are often intentionally mislabelled or stated as being grown from domestic stock when in fact they were taken from the wild. Correct identification requires the services of a plant taxonomist, whereas wild-collected specimens can be identified by damage caused when the plant was dug up and the typical 'wear and tear' damage that all wild plants receive, such as porcupine teeth marks in the case of South African plants. A plant's bulb and roots would also retain adherent samples of the soil from which it was obtained. This would facilitate stable isotope analysis, although this would require access to suitable reference materials. However, even basic soil analysis would demonstrate whether the plant had been grown in the place it was claimed to come from.

In Britain, all native plants receive protection under the Wildlife and Countryside Act (1981), which states that it is illegal to dig up any wild plant without appropriate permission and some species, such as bluebells and snowdrops, are also protected from being picked and sold – although this sometimes happens on a large scale and is conducted as an organized criminal activity. The Act also covers algae, mosses, lichen and fungi.

Quick quiz

1. In which one of the following body parts would the presence of diatoms be the best indication of death by drowning: nasal sinuses, stomach, lungs, bone marrow? Explain your reasons.

2. Why does extreme care need taking when sampling for diatoms and interpreting forensic evidence based upon them?

3. In relation to the 'magic mushroom' *Psilocybe semilanceata*, which of the following practices is illegal: picking wild mushrooms, selling wild mushrooms, selling dried wild mushrooms, selling frozen cultivated mushrooms, selling food containing cooked wild mushrooms?

4. How can an examination of tree rings be used to demonstrate whether a church panel painting is a forgery or genuine?

5. State two ways in which the growth of roots might be used to estimate the length of time a corpse has been buried?

6. What is a palynomorph? Give two examples of palynomorphs.

7. How can palynomorphs be used to demonstrate that someone who died and was buried in one part of the country was subsequently exhumed and buried again elsewhere?

8. Why should people be prevented from leaving floral tributes near the site of a murder until after the forensic team have finished collecting their samples?

9. How can stable isotope analysis be used to demonstrate the provenance of drugs and foodstuffs?

10. How is it possible to distinguish between cycads grown from domestic stock and those collected illegally in the wild?

Project work

Title

Can diatoms enter the bloodstream via the food or drinking water?

Rationale

One of the diagnoses of drowning is the discovery of diatoms in the lungs and bone marrow. However, there is limited published evidence to preclude the possibility that diatoms might also enter the body via being breathed in or through the gut.

Method

Laboratory rats or mice would be fed a diet supplemented with diatoms either in the food or the drinking water. Freshwater diatoms can be cultured relatively easily and might also be radioactively labelled to aid finding them again. After allowing the animals to feed for known amounts of time they would be sacrificed and the body organs analysed for the presence of diatoms.

Title

Determining the duration of submergence from algal colonization.

Rationale

Criminals often dispose of the objects that would link them with a crime – such as a weapon, or an empty wallet – in a river or canal. It would therefore sometimes be useful to know how long an object retrieved from the water has been there.

Method

Objects of varying textures and chemical and physical properties, such as knives, screwdrivers and leather and plastic wallets, would be placed at different depths in a watercourse at different times of year. They would be retrieved after set periods and the extent of the algal colonization established.

Title

Determining the duration an object has rested on the ground from changes in the underlying vegetation.

Rationale

It is sometimes useful to know how long a body or an object has been resting at a particular location.

Method

Organic (e.g. a body or piece of meat, clothed or unclothed) and inorganic (e.g. knives, baseball bats, broken glass bottles) objects would be placed on the ground at varying times of the year. A uniform surface, such as a lawn or grassy field, would make life easier for preliminary experiments. After set periods of time, the objects would be removed and changes in the underlying vegetation noted. This could be done by noting changes in coloration (the levels of chlorophyll could be measured spectrophotometrically), growth characteristics (e.g. internodal length, leaf area), and plant chemistry (e.g. levels of starch, sugars and chlorophyll) and comparing them to surrounding uncovered plants.

Title

The effect of burial on root growth.

Rationale

When a body is buried, plant roots will be damaged but growth may also be stimulated by the release of nitrogen and organic matter as the body decays.

Method

This would be best done in a location where minor damage to tree roots would be caused and compare the effect on root growth of burying an organic and an inorganic object. When the objects are buried, photographic evidence should be taken of the damage to the roots and as roots do not grow continually and uniformly throughout the year the experiment should be repeated in different seasons. After set periods of time, the buried objects would be carefully retrieved and a record made of the extent of the root growth – root lengths, branching, diameter – towards, into and around the buried object.

A simple demonstration of the effects of a dead body on both above- and below-ground plant growth can be achieved by burying meat at varying depths within a glass-fronted 'wormarium' in which grasses are growing. There are many variations that can be made, such as introducing earthworms, burning the meat to varying extents and recording everything on time-lapse photography or video recording.

6 Invertebrates in forensic science

Chapter outline

Invertebrates as forensic indicators in cases of murder or suspicious death
  Invertebrates attracted to dead bodies: detritivores; carnivores; parasitoid insects; coprophiles
  Invertebrates leaving dead bodies
  Invertebrates accidentally associated with dead bodies
  Invertebrates as a cause of death
Invertebrates as forensic indicators in cases of neglect and animal welfare
The role of invertebrates in food spoilage and hygiene litigation
Invertebrates as a cause of nuisance
Invertebrates as a cause of structural damage
Illegal trade in protected species of invertebrates

Objectives

Review the biology of the most important UK species of invertebrate detritivores and how they sequentially exploit a dead body as a food source.

Describe the types of predators and parasitoids likely to be found associated with a dead body and their impact upon the detritivore community.

Discuss the factors that attract coprophilic invertebrates to a dead body.

Describe why ectoparasitic invertebrates leave a dead body and some species of invertebrate become accidentally associated with one.

Explain how invertebrates may be either a direct or indirect cause of human fatalities.

Explain how invertebrates may become involved in cases of neglect and animal welfare.

Discuss the role of invertebrates in food spoilage and hygiene litigation.

Discuss the role of invertebrates as a cause of nuisance and structural damage.

Discuss the illegal trade in protected invertebrate species.

Essential Forensic Biology, Alan Gunn
© 2006 John Wiley & Sons, Ltd

Introduction

Invertebrates are metazoan animals that lack a backbone and include creatures as diverse as mites smaller than the nucleus of a large protozoan to giant squid several metres long. The arthropods are but one group of invertebrates but they are the most successful in terms of their abundance, biomass and numbers of species. Arthropods are characterized by their possession of a hardened exoskeleton that is shed periodically to accommodate growth and they have specialized jointed appendages. Typical examples are scorpions, crabs and moths. It is the arthropods, and in particular the insects, that are the most common sources of forensic evidence. The study of insects is referred to as 'entomology', a word derived from the Greek words *entomon* (an insect) and *logos* (science). Logic would therefore suggest that forensic entomology is the use of insects in legal investigations. However, many workers restrict the definition to the use of insects in murder and suspicious death investigations, thereby excluding the numerous other legal cases in which insects are involved. Sometimes, the term 'medicocriminal entomology' is used to emphasize this restricted definition.

Insects account for 72 per cent of all known species of animals and together with other terrestrial arthropods they can be found in virtually every terrestrial habitat, both natural and man-made, from the polar regions to the equator. Within these regions, insects and other arthropods occupy virtually every conceivable ecological niche. Some are specialist herbivores, carnivores, detritivores or parasites, whilst others are generalists that consume a variety of foodstuffs. Their lifestyles sometimes bring them into conflict with mankind because they consume the same food as ourselves, use us as their food or occupy our dwellings. Knowledge of their biology is therefore useful in reducing the harm that certain species may cause either directly or indirectly (e.g. through spreading diseases) and in understanding how they can provide forensic evidence. However, the vast majority of arthropods and other invertebrates are harmless to mankind: many are extremely valuable as food, as pollinators of our crops or as biological control agents and they all have an important role in the normal functioning of ecosystems.

Invertebrates as forensic indicators in cases of murder or suspicious death

The invertebrate species that provide forensic evidence in cases of murder or suspicious death investigations can be broadly grouped into those that are attracted to dead bodies, those that leave dead bodies and those that become accidentally associated with the dead body and/or the crime scene.

Invertebrates attracted to dead bodies

Numerous invertebrate species are attracted to dead bodies, including detritivores that feed on the decaying tissues, carnivores and parasitoids that come to prey on the detritivores and coprophiles that tend to feed on faeces rather than the decaying tissues of the dead body.

Detritivores

A dead body represents a temporary source of easily metabolized organic matter that, unlike that of living organisms, is chemically and physically unprotected. In the initial stages, it therefore attracts highly mobile r-type species that are adapted for finding and exploiting such temporary food sources. These r-type species are typified by short life cycles and the ability to undergo explosive increases in population under the right conditions. They rapidly consume the most easily degradable tissues until all that remains is material that does not have sufficient nutritive value to support their growth. Therefore, with time, the r-type detritivore species are replaced by species with longer lifecycles and lower reproductive rates that are able to exploit food that is more difficult to metabolize and has a lower nutritive value.

Flies (order Diptera), especially blowflies, are usually the first organisms to arrive at a corpse (Figure 6.1), sometimes within minutes of death, and they are also the species of greatest forensic importance (e.g. Goff 2000; Byrd and Castner, 2001; Arnaldos, *et al.*, 2005). Blowflies belong to the family Cal-

Figure 6.1 Blowflies ovipositing within and around the nasal cavity of a sheep. The flies began egg-laying within 5 minutes of the animal being killed. In addition to the body, egg batches may also be laid upon the vegetation underneath a corpse

liphoridae and are commonly called bluebottles or greenbottles. Numerous species of blowfly exist and it is not unusual to find several species on a single corpse. However, not all blowfly species utilize corpses, and corpses may contain the larvae of many different species of Diptera other than those belonging to the family Calliphoridae. The common name 'blowfly' is derived from the noun 'blow', which means a mass of fly eggs. Food, a wound or a corpse covered in fly eggs is therefore said to be 'fly-blown' and the insect responsible is a 'blowfly'. The common names 'bluebottle' and 'greenbottle' may be derived from the Gaelic '*boiteag*' meaning a 'maggot'. If this is correct, a greenbottle is therefore a green fly that produces maggots (Erzinclioglu, 1996). Because many flies look superficially similar, confusions may arise so it is better to use proper taxonomic terms rather than words such as 'greenbottles', etc.

The life cycle of the 'typical' blowfly species, such as those found on dead bodies, is relatively straightforward. The gravid female fly (i.e. one ready to lay) lays her eggs on the corpse. She usually chooses one of the natural openings, such as the nose, ears and mouth, the eyes or the site of a wound. The anus and genitalia may be used if these are accessible and eggs may also be laid on the vegetation underneath a corpse. Blowflies usually colonize corpses that are still fresh or during early stages of decay and are unlikely to lay their eggs on one that has started to dry out and skeletonize – although the adults may feed on a corpse at this stage. Eggs are laid in batches that may number up to 180 and a female fly may lay a total of several thousand eggs over the course of her life. When an egg hatches a first instar larva emerges. At this stage it is small and delicate (Figure 6.2) and it rapidly moves to where it can find optimum condi-

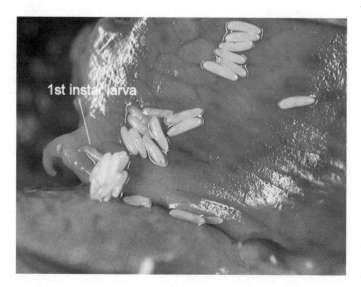

Figure 6.2 Mature blowfly eggs about to hatch. The developing larvae can be seen within the egg cases

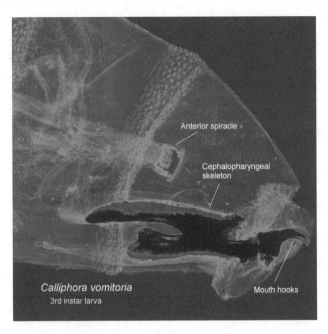

Figure 6.3   Anterior of the third instar larva of the blowfly *Calliphora vomitoria*. The cephaloskele-ton has a large surface area for the attachment of muscles, thereby allowing the mouth hooks to be dragged back and forth with considerable force

tions – not too damp and not too dry – and starts to feed. Blowfly larvae feed using chitinous mouth-hooks that drag material into their oral cavity (Figure 6.3). They feed on both the tissues of the corpse and the bacteria that grow on it. In addition to the physical act of feeding, maggots also release enzymes and other substances that help to break down the underlying substrate. After approximately 24–48 hours the larva moults to the second instar, and after a further 24–48 hours the larva moults to the third instar. The third instar larva feeds voraciously and rapidly increases in size and weight over 3–4 days (Figure 6.4). Development times are strongly influenced by temperature and some blowfly species develop much faster than others. Once the third instar larva has developed sufficiently, it empties its gut contents and (usually) leaves the corpse in search of somewhere to pupate. These maggots can be distinguished from those that are actively feeding because their gut is clear rather than dark – this can be useful in determining the age of a larva. Pupation usually takes place in the soil and its duration, like all the other stages, is highly temperature depend-ent and may last from just a few days to several months. As a maggot begins to pupate, its body contracts and when it moults the third instar cuticle is retained to form a protective puparium surrounding the pupa. The larvae of species such as *Calliphora vicina* may travel several metres from the corpse in search of a pupation site but others, such as *Protophormia terraenovae*, seldom move far, and may even pupate among the clothing associated with the body

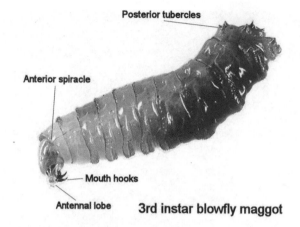

Figure 6.4 Typical third instar blowfly larva. The full crop indicates that this larva has not yet finished feeding

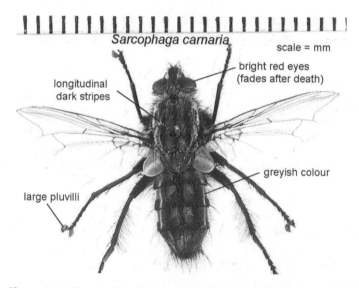

Figure 6.5 Characteristic features of an adult fleshfly (Sarcophagidae)

(Nuorteva, 1977; Smith, 1986). After the adults emerge, they fly off in search of food. In addition to carrion, blowflies also feed on nectar (they can be important pollinators of flowers), honeydew, dung, rotting fruit and other decomposing matter. In many species, the female flies need a protein meal before they can lay their eggs, although this may not be necessary for the males to mature their sperm. It is said that the female blowflies typically only mate once (this is unusual in all animals) although the male may mate many times (Erzinclioglu, 1996). The adult lifespan varies between species and can be from a few weeks to several months. Those species that overwinter as adults may live for several months at this time of year although adults of the same species of fly emerging during the spring or summer may have a much shorter lifespan. Most lifespan studies involve laboratory-reared insects and this can result in overestimates because in the wild the majority of flies would die from predation, starvation and disease rather than reaching their maximum potential age.

Sarcophagid flies (Figure 6.5), commonly known as fleshflies, often compete with blowflies during the early stages of decomposition. Some workers state that they are among the first insects to arrive at a corpse whilst others state that they tend to 'arrive after the main blowfly sequence'. The discrepancy probably results from differences in biology between the various species of *Sarcophaga* and the individual circumstance. Adult *Sarcophaga* species are reportedly more willing to fly during wet weather than blowflies and may, therefore, be among the first to colonize a corpse under these conditions (Byrd and Castner, 2001). The adult flies are usually greyish in colour with longitudinal dark stripes on their thorax and bright red eyes, although the red colour fades after death. The tarsal claws and the pluvilli (tarsal pads) are usually large and give the fly an appearance of being 'big-footed'. The eggs hatch in the female fly's reproductive tract and she therefore lays first instar larvae. In addition to corpses, the larvae of various species of *Sarcophaga* are also found in dead organic matter and faeces, etc. The larvae can be distinguished from those of blowflies, etc. by their spiracles, which are partially hidden in a deep cavity, but they are otherwise very difficult to identify.

In addition to the blowflies and the fleshflies, numerous other fly species can be found on corpses although most of these tend to arrive in the later stages of decay and their forensic potential is more limited. Examples include the phorids, stratiomyids, piophilids, syrphids and trichocerids. Phorid flies (Figure 6.6) belong to the family Phoridae – a large group that contains over 2500 species and examples of virtually every lifestyle (Disney, 1994). They are all small flies, 1.5–6.0 mm in length, and have a characteristic 'humped' profile. The larvae of many species are detritivores and some of these, of which the coffin fly, *Conicera tibialis*, is the most well known, have been found in dead bodies. The coffin fly gets its name from the ability of the adult flies to detect corpses buried over a metre below ground and then to crawl down through cracks in the soil to reach them. Successive generations of coffin flies can occur totally underground on a buried corpse although it may also be found on bodies left above ground. Other phorid flies that may also be found on corpses include *Diplonerva florescens*, *Megaselia abdita*, *Megaselia rufipes*, *Megaselia scalaris* and *Triphleba hyalinata* and they all tend to be found during the later stages of decay as the body is starting to dry out. Owing to their small size, phorid flies are often able to gain access to containers (such as coffins) and rooms that would exclude blowflies.

Syrphid flies (family Syrphidae) are commonly known as 'hover-flies' and they are often associated with flowers, sap-runs and other sugary foods. Their larvae exhibit a wide range of feeding strategies, including predation, herbivory and detritivory. Among the detritivores, the larvae of *Eristalis tenax* (Drone fly) are the ones most likely to be encountered on a dead body, especially during the middle stages of decay when it has started to liquify. Commonly known as 'rat-

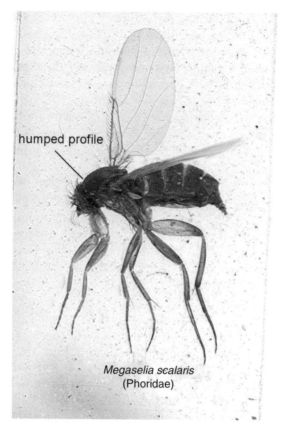

humped profile

Megaselia scalaris
(Phoridae)

Figure 6.6 Characteristic features of an adult Phorid fly

tailed maggots', the larvae have a long telescopic breathing tube that acts like a snorkel and enables them to breathe whilst the rest of their body is submerged (Figure 6.7). However, when they have finished feeding, the larvae crawl away in search of a drier environment in which to pupate.

Piophilid flies are usually associated with corpses in the later stages of decay, when the body is starting to dry out, although there are records of them appearing earlier. The adult flies are small and shiny-black, which in a certain light gives a bright green or blue sheen. *Piophila casei* (Cheese skipper) is the best known of the piophilids, as it is also a common pest of stored products, although several other species may also be found on corpses. Both *Piophila casei* and *P. foveolata* have been associated with human corpses.

Stratiomyid fly larvae are common soil invertebrates and they are often found on buried bodies or underneath bodies left on the soil surface. They have a characteristic flattened profile (Figure 6.8) and a long conical head whilst their surface has a grained, leathery appearance as a consequence of the deposition of calcium carbonate crystals in the cuticle. The adults are called 'soldier flies'

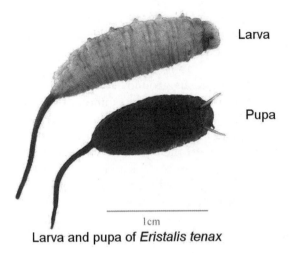

Larva

Pupa

1cm

Larva and pupa of *Eristalis tenax*

Figure 6.7 Larva and pupa of the hoverfly *Eristalis tenax*. The telescopic breathing tube is a characteristic feature of this genus

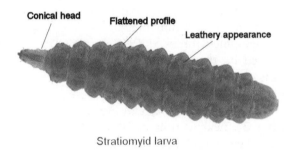

Stratiomyid larva

Figure 6.8 Characteristic features of a stratiomyid larva. These are common soil organisms but can be very numerous within bones and the skull once a dead body has become skeletonized

because, in many species, they are brightly coloured and their scutellum (a shield-shaped plate on the dorsal surface of the thorax) is armed with backwardly pointing spines. Stratiomyids do not tend to be used in forensic studies in the UK although in America *Hermetia illucens,* the 'black soldier-fly', has been used to estimate time since death in the absence of data from blowflies (Lord *et al.*, 1994). Female *H. illucens* are usually considered to begin laying their eggs on a corpse once it is 20–30 days old and entering the advanced decay or putrid dry remains stage of decomposition. At this stage, blowfly maggot activity has declined or ceased entirely and in their place *H. illucens* larvae may become the dominant insect fauna on the corpse. However, in some circumstances it may start to colonize a corpse that is less than 7 days old and therefore still relatively fresh (Tomberlin *et al.*, 2005): this clearly has implications for any estimations of the minimum time since death that are based primarily on the stage of development of this species. *Hermetia illucens* is not found in the UK although it has been recorded in Continental Europe.

Trichocerid flies are commonly called 'winter gnats' because adult mating swarms may be seen dancing above lawns, fields and hedgerows during the winter, even when snow is on the ground. The adult flies are also to be seen during the spring and autumn but are less common during the summer months. There are only ten species in the UK and the adults resemble small crane flies (daddy-long-legs). The larvae are long, thin and cylindrical with a well-developed head and a number of them are known to be detritivores. *Trichocera saltator* has been found on human corpses and they are most likely to be found during the winter period when blowfly activity is reduced.

The beetles, order Coleoptera, comprise 25 per cent of all animal species described to date and they are found in a wide diversity of habitats and exhibit a range of lifestyles – including the colonization of dead bodies. Dermestid beetles (family Dermestidae), commonly called larder beetles, hide beetles, carpet beetles or leather beetles (Figure 6.9), and their larvae are not found on fresh bodies but can quickly skeletonize it once it has started to dry out (Bourel *et al.*, 2000; Schroeder *et al.*, 2002). They are small, usually less than 10 mm long, and are common household pests that feed on organic matter in carpets, etc. There are several species, although their life cycles and habits are fairly similar and their development times and the numbers of generations per year are strongly influenced by the environmental conditions. In the case of *Dermestes lardarius* the gravid female lays eggs directly on the food source in batches of six to eight over a 3 month period, these hatch after 7–8 days and five to seven larval instars follow. The final instar larva leaves the food source and pupates in a burrow nearby after which the adult insect emerges – this is 2–3 months after the eggs were laid. Adult dermestid beetles are often cannibalistic on their own eggs and larvae and should therefore be kept separate from these when collecting specimens (Archer and Elgar, 1998).

The caterpillars (larvae) of several species of pyralid and tineid moths, order Lepidoptera, are sometimes associated with corpses. For example, those of *Tinea bisselliella* and *Tinea pellionella* are common household pests that normally feed on stored clothing and carpets, stored grain, cereals and dry vegetable matter but they may also be found on corpses when the body has become dried and skeletonized. They are even able to break down the keratin found in hair and fingernails. Similarly, the larvae of the pyralid moths *Aglossa caprealis* and *Aglossa pinguinalis* may be found on corpses that have dried out, although their normal environment would be on stored agricultural produce and decaying vegetation. A gravid female *Tinea pellionella* typically lays 40–70 eggs over approximately 24 days. The female moths do not fly actively whilst they are laying their eggs. Moths seen flying around a room are therefore usually males or those females that have finished laying eggs. The eggs hatch in 7–37 days depending on temperature and the larvae often spin a protective silken tube around their bodies. The larval period may last anything from 2 months to 4

Dermestes lardarius

(a)

(b)

Figure 6.9 Adult dermestid beetle (*Dermestes lardarius*) (a) and an unidentified larval dermestid beetle (b). The larvae and adults of these beetles are seldom found on corpses until the remains have begun to dry out but they are common household and stored product pests

years, depending on temperature, humidity and the nature of their diet, after which the mature larva pupates within its protective tube and the adult moth emerges after 11–54 days.

Mites (Acari) are frequently abundant on corpses, the species composition changing as the body goes through its varying stages of decay. For example, gamasid mites (e.g. *Macrocheles*) tend to be abundant during the early stages of decay, and tyroglyphid and oribatid mites (e.g. *Rostrozetes*) are more numerous once the body has started to dry out. Some of the mite species feed on the corpse, others are predatory and feed on the detritivores, etc. and some feed on the fungi, bacteria, etc. that grow on the body. Investigating the potential of Acari as forensic indicators has been hampered by their small size and the scarcity of mite taxonomists. Consequently, there is little published information on the mite fauna of corpses and mites are not used to determine time since death.

'Worms' are popularly associated with dead bodies. For example, one of the lines in the folk song 'On Ilkley Moor Baht'at' goes 'then t'worms 'll cum and eat thee oop'. However, the worms referred to in the song and in popular folklore are almost certainly the maggots of blowflies – anything small white and wriggly tends to get classed as 'a worm'. Earthworms are, of course, a quite distinct group of organisms and there are about 26 species (plus some introduced species with a restricted distribution) in the UK (Sims and Gerard, 1985). Although earthworms will feed on decaying bodies, they are sensitive to changes in pH and would be repelled by the seepage of large amounts of acidic material into the soil. Earthworms, especially epigeic species (those living on the soil surface), may enter buildings so their presence on a corpse found indoors does not mean that it was previously buried or left lying on the surface of soil. Earthworms decay rapidly to a foul-smelling mush once they are dead so, if a body containing earthworms is disposed of in the sea or a lake, they are unlikely to provide evidence of the body's previous location unless the body was found very quickly. Earthworms accumulate heavy metals, including arsenic, but it is not known whether they accumulate other toxins from a dead body.

Many other soil invertebrates, such as nematodes, slugs, snails, Collembola, Diplura and millipedes, may be found on corpses, especially those that are buried or left on damp soil, but they are seldom reported as being helpful in forensic studies. This is partly a consequence of the difficulties associated with identification and partly because in many cases they develop much more slowly than blowfly larvae and are therefore less useful in determining the passage of time. However, monitoring the changes in the abundance and diversity of nematodes and Collembola has proved useful in studies on pollution ecology and the appearance of a large decaying body is, in some respects, a pollution incident, so their forensic potential might be worth investigating further. Similarly, snails accumulate heavy metals but it is not known whether they would also accumulate other toxins from a dead body. Snail shells are slow to decay and may last many years under suitable conditions so their potential for yielding forensic evidence should be considered.

Carnivores

The rapid development of large numbers of maggots and other detritivores on a corpse soon attracts the attention of both carnivores and parasitoids. Some of the carnivores will also consume dead organic matter, so the distinction between predator and detritivore is not always a simple one. Many beetles, especially certain carabid and staphylinid species, are important predators of other soil invertebrates during both their adult and larval stages. Some species of carabid beetle have even been used as biological control agents but their impact on the corpse fauna is not known. The larvae of many carabid and staphylinid beetles are very similar in appearance (Figure 6.10). The carabid larvae can be distinguished from the shape of their legs. The carabid beetle larvae have six-segmented legs whilst those of staphylinid larvae are five-segmented. In addition, the legs of many species of carabid end in two claws, whilst those of staphylinids have only one claw. Histerid beetles tend to arrive on corpses several days after blowflies and, during daylight hours, they are usually found underneath the body. The majority of species are predatory as both larvae and adults, although they also consume dead organic matter. Some species consume

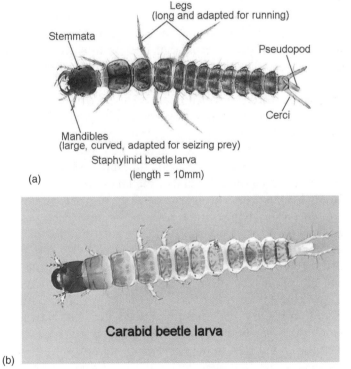

Figure 6.10 Larvae of a staphylinid beetle (a) and a carabid beetle (b). Although fearsome predators, their impact on the development of a maggot population is not known

blowfly eggs and larvae although the consequences of this for the establishment of the blowfly population and the rate of decomposition are not known.

Burying beetles (family Silphidae) are so called because the adults of several species bury dead animals, such as mice, upon which their larvae subsequently feed. They are capable of detecting a dead body at a considerable distance and are one of the first invertebrates to arrive at a fresh corpse. In those species that bury dead bodies, several individuals will cooperate in the burial and then fight to determine which one will lay her eggs on the corpse. They are incapable of burying an entire human corpse but may inter bits if it has been dismembered. Alternatively, they may chew out hunks of tissue, which are then buried. After burying the corpse or body part, the female beetle lays her eggs above the carcass and then covers it with her hind gut secretions to inhibit the growth of bacteria and fungi. The eggs hatch after about 2.5–5 days and the female remains with the hatchlings. In some species, the male also remains and helps rear the young. Depending on the species, the adult(s) may feed the young directly from a depression they make at the top of the carcass. Larval development takes 5–8 days. After the final instar, the larvae crawl off and pupate for about 2 weeks. The adult beetles then emerge but they do not reproduce until the following year. Not all burying beetles actually bury corpses, e.g., *Necrodes littoralis* usually breeds within the bodies of large vertebrates and has been recorded entering via stab wounds. Both adult and larval burying beetles will consume blowfly maggots and some burying beetle species act as transport hosts for mites belonging to the family *Poecilochirus* that feed on blowfly eggs. There is therefore a theoretical possibility that burying beetles may influence blowfly maggot numbers and age distribution.

Many ant species (order Hymenoptera) are predatory and if a corpse is located close to their nest they may slow the establishment of the blowfly population through removing eggs and young larvae. However, they may also feed on the tissues of the corpse and thereby speed the process of decay. Indeed, one of the recommended methods of preparing an animal skeleton for display used to be to bury it in an ants' nest. Similarly, the larvae of all social wasps belonging to the family Vespidae are carnivorous and the adult wasps prey on a variety of other invertebrates that they bring back to the nest. A large wasp nest may therefore be expected to have an impact on the surrounding invertebrate fauna (Archer and Elgar, 2003) whilst the adult wasps will also forage for scraps of meat from nearby corpses. Not all ant species are carnivorous and therefore likely to have an impact on the rate of decomposition. It is therefore important to ensure a correct identification of any ants found in large numbers around a dead body. Goff and Win (1997) have described a case in Hawaii in which the postmortem interval for a body discovered in a metal trunk was estimated from the time taken for an ant nest to develop within the trunk to the stage at which alate (winged) reproductive castes were being produced.

All centipedes (Chilopoda) and spiders (Aranea) are predatory and although they are often found on corpses their impact on the other fauna is not known.

However, considering the enormous reproductive potential, rapid growth rates and powers of recruitment of the typical blowfly species, it is unlikely that they exert a significant effect. Certain centipedes are found naturally in houses and outbuildings, whilst others – such as the long, thin geophilomorph species – are usually only found in the soil. Although spiders are commonly associated with spinning webs, many species actively hunt or stand guard at the entrance to their refuge and rush out to grab their prey as it comes past. A variety of species of spider may be found on corpses, sometimes because they are preying on the other corpse fauna but also because they are using it as a refuge or as an anchor for their web.

Parasitoid insects

Parasitoid insects are those that lay their eggs within the bodies of other invertebrates, usually other insects. The eggs hatch within their host, but this is not killed until the parasitoid has completed its larval development. Examples of parasitoid wasps attacking blowfly and housefly larvae or pupae include *Nasonia vitripennis, Alysia manducator, Muscidifurax raptor* and *Spalangia cameroni*. This can result in the slowing of the larval or pupal development rate and ultimately the death of the parasitized insect. *Nasonia vitripennis* lays its eggs in the pupae of a range of blowfly species. The adult female wasp (Figure 6.11) uses her ovipositor to bore a hole through the puparium and lays her eggs on top of the developing pupa. This occurs 24–30 hours after the pupa has formed, i.e. the point at which the third instar larval cuticle separates from the pupal cuticle and forms the puparium. After hatching, the wasp larvae feed on the fly pupa, killing it in the process. The wasps pupate within their host's pupar-

Figure 6.11 Adult parasitoid wasp *Nasonia vitripennis*. This species of parasitoid lays its eggs within the pupal cases of blowflies but other parasitoid wasps lay their eggs within the larvae and are often seen searching for victims from within a maggot feeding ball

ium and the adults chew their way out after 10–50 days, depending on temperature. *Alysia manducator* has a similar life cycle but it lays its eggs directly inside the developing pupa. The forensic importance of parasitoids is that they can be locally abundant and therefore if blowfly larvae or pupae are being collected for rearing purposes a large sample size should be obtained to allow for parasitoid-induced mortality. Because *Nasonia vitripennis* only lays its eggs on fly pupae 24–30 hours old, recording when the adult parasitoids emerge can be used to determine the postmortem interval (Grassberger and Frank, 2003). This is because if one knows how long the wasps take to develop it can be calculated when they laid their eggs and on that date the pupae would be 24–30 hours old, from which point it would be possible to determine when the blowfly eggs were laid and therefore the approximate date on which the person died. The possibility of using parasitoids of phorid flies in a similar manner has also been suggested (Disney and Munk, 2004).

Coprophiles

Although there are specialist coprophilic species for which faeces is their only food, many also feed on other sources of decaying matter. For many invertebrate species, therefore, coprophagy and detritivory are simply options that are indulged in to varying degrees. Coprophilic species might be present when a body is coated with faeces (murderers sometimes desecrate their victim's body this way), when the gut contents are exposed through wounds or the decay process and when fouling of the clothing occurred before death, possibly through neglect or at the time of death through fear or the manner of death. For example, strangulation and suffocation often results in the victim losing control of their bladder and anal sphincters during their death struggle.

A number of species of muscid flies are commonly associated with faeces although there are over 450 species in the UK and as a group they exhibit a wide variety of lifestyles at both the adult and larval stages (Skidmore, 1985). Some species, such as *Musca domestica* (House fly), *Fannia canicularis* (Lesser housefly) and *Fannia scalaris* (Latrine fly), have adults that readily enter buildings and larvae that live in faeces or dead organic matter. Consequently, these species are common in unsanitary conditions and will infest soiled clothing, even if the wearer is still alive. They may be found on corpses, especially if it is soiled with faeces or the gut contents are exposed. They are usually found at the later stages of decay and are not usually present on a fresh corpse. *Musca domestica* prefers relatively high temperatures and, outdoors, the adults usually do not appear until May–June. However, indoors they will breed continuously wherever the temperature remains high enough. The female flies lay their eggs in batches of about 150 eggs within crevices of the larval food medium. The eggs can be difficult to see when laid on white material such as nappies or bandages. The larvae develop rapidly and the whole life cycle from egg-laying to adult emergence may take as little as 6–8 days under optimal conditions. Conse-

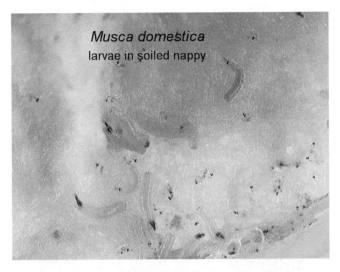

Figure 6.12   Larvae of the common housefly *Musca domestica* developing within a soiled nappy

quently, populations may build up extremely rapidly. The larvae do not, as a rule, infect wounds in living humans or other animals, although they may be found in blood-soaked clothing and hair. They are often found in clothing soiled by urine or faeces (Figure 6.12) and this infestation may begin before the wearer died. This is important to bear in mind when calculating the time since death.

Invertebrates leaving dead bodies

Fleas, body lice and head lice leave their host soon after it dies and the body temperature starts to decline. The sight of them crawling on top of a dead person's clothing is therefore an indication that he/she has not been dead for long. The so-called human flea, *Pulex irritans*, is no longer a common pest in the UK and people are more likely to come into contact with the cat flea, *Ctenocephalides felis*, and the dog flea, *Ctenocephalides canis*. Humans may harbour one or two fleas, usually acquired from their pets, and large numbers are only present in people suffering from neglect or living in unsanitary conditions. The same is true of the body louse *Pediculus humanus humanus*, but the head louse *P. humanus capitis* (Figure 6.13) is extremely common, especially in children, even among affluent families. Crab lice (Figure 6.14) are relatively common among the sexually promiscuous and those who have a relationship with someone who is. Although they are also referred to as 'pubic lice', they may also be found on other coarse body hair, such as beards. They move very slowly and are therefore less likely to be seen wandering away from a dead body. The potential of ectoparasites such as fleas and lice to act as a vector of disease needs to be borne in mind by persons working with dead bodies.

Figure 6.13   Human head louse, *Pediculus humanus humanus*

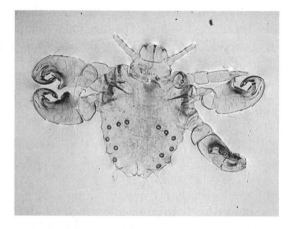

Figure 6.14   Human crab (pubic) louse, *Pthirus pubis*. These lice may also be found on the beard, eyelashes and other coarse hair

Invertebrates accidentally associated with dead bodies

Invertebrates that are not associated with the decay process may become accidentally associated with a body (see Table 6.1) through becoming trapped in its clothing or from using the corpse as a refuge. Alternatively, they may become trapped inside the vehicle or container in which the body is found. If these invertebrates have a restricted geographical distribution, are associated with very spe-

Table 6.1 Summary of insects commonly associated with dead bodies

Insect	Association
Diptera	
Blowfly adults	Lay eggs on corpse during early stages of decay
	Feed on corpse at all stages of decay
Blowfly larvae	Feed on corpse during early stages of decay
Fleshfly adults	Lay eggs on corpse during early stages of decay
	Feed on corpse at all stages of decay
Fleshfly larvae	Feed on corpse during early stages of decay
Muscid adults	Lay eggs on soiled clothing and faeces
	Feed on corpse at all stages of decay
Muscid larvae	Feed on faeces and soiled regions, also found at later stages of putrefaction when corpse is still moist but abundance of blowfly larvae declining
Phorid larvae	Feed on corpse during later stages of decay. Can colonize buried corpse
Stratiomyid larvae	Feed on corpse once body has dried out
Piophilid larvae	Feed on corpse during later stages of decay
Eristalid larvae	Feed on soiled and liquefied regions
Nematocera larvae	Feed on corpse during colder months when blowflies are inactive or their numbers are low
Coleoptera	
Carabidae	Predators as adults and larvae. Feed on other insects and on corpse
Staphylinidae	Predators as adults and larvae. Feed on other insects and on corpse
Histeridae	Predators as adults and larvae. Feed on other insects and on corpse. Arrive a few days after the blowflies
Silphidae	Predators as adults and larvae. Feed on other insects and on corpse. Adults arrive during early stages of decay.
Dermestidae	Detritivores as adults and larvae. Feed on corpse once it has dried out
Hymenoptera	
Vespid wasps	Adults predatory on other insects. Will remove flesh from corpse
Parasitoid wasps	Adults lay eggs in larvae or pupae of Diptera
Ants	Some species predatory on other insects. Will remove flesh from corpse
Lepidoptera	
Pyralid moths	Larvae feed on corpse once it has dried out
Tineid moths	Larvae feed on corpse once it has dried out
Blattaria	
Cockroaches	Only found indoors in UK. Nymphs and adults will feed on corpse at all stages of decay
Anoplura	
Head lice, body lice	Parasites that leave dead body as it starts to cool
Collembola	
Springtails	General detritivores found underneath bodies left on soil, especially at later stages of decay
Diplura	General detritivores found underneath bodies left on soil, especially at later stages of decay

cific habitats or are active at only specific times of year this can provide evidence of a person's association with a particular locality at a particular time.

Invertebrates as a cause of death

Within the UK, few native invertebrate species have venoms capable of causing a serious threat to human health. However, some may cause death by inducing an anaphylactic reaction in sensitive individuals or through stinging a person in the mouth or throat – this can result in rapid swelling that can cause asphyxiation. Invertebrates may also cause human deaths indirectly by causing distraction or panic that results in a fatal accident. In the UK and northern Europe most fatalities involving invertebrates are associated with honeybee or wasp stings and there are an average of two to nine cases per year in the UK. Honeybees leave their sting behind after it has been used and this results in the bee dying. A bee sting can therefore be diagnosed from the physical presence of the sting or from finding the bee's dead body. Wasps and hornets do not leave their sting and unless the fatal event was witnessed their involvement may only be indicated by the presence of a nearby nest. The fashion for keeping exotic pets has led some persons to keep potentially dangerous invertebrates such as scorpions, certain species of spider and large tropical centipedes. The practice is undoubtedly risky and the more dangerous species are to be covered by legislation requiring owners to obtain a license and to keep their animals under strictly controlled conditions.

Spiders and insects such as wasps and bees can induce panic attacks in people with a phobia for them. They are therefore thought to be involved in a number of otherwise 'unexplained' vehicle accidents. A wasp or bee can enter the vehicle via a window, for example, and the driver subsequently loses control in an attempt to swat or avoid it. The driver need not be stung and the insect will almost certainly fly away unseen after the event. Should the accident cause the death of the driver, even if he or she was stung, there may not have been time for an allergic reaction to develop and any sting–associated pathology could be missed at autopsy. Similarly, a person may run into fast-moving traffic in an effort to escape from a buzzing insect or a spider that has become entangled in their hair. Bumblebees are highly unlikely to sting but may induce panic reactions owing to their large size and loud buzzing behaviour. Cyclists can suffer serious eye injury as a consequence of insects getting into their eyes. If they lose control of their bikes they may suffer fatal injuries.

Allergens present in the faeces and chitin of some invertebrates may induce strong reactions in sensitive individuals. For example, the house dust mite, *Dermatophagoides pteronyssinus*, is a common cause of allergy and can cause asthmatic attacks and other symptoms. Recent improvements in the ability to diagnose anaphylaxis at postmortem have indicated that some cases in which persons suddenly died of unknown causes might be linked to hypersensitivity to this mite (Edston and Van Hage-Hamsten, 2003). Similarly, locusts can induce potentially fatal reactions in sensitive individuals. Animal technicians who work with locusts on a regular basis must therefore take appropriate precautions. In countries where locusts occur naturally, human illness has been associated with their swarming behaviour. For example, during September 2003 in a town in central Sudan, eleven people were reported dead and over a thousand hospitalized with breathing difficulties as a consequence of asthmatic reactions to the sudden influx of large numbers of locusts.

Invertebrates as forensic indicators in cases of neglect and animal welfare

The invertebrates involved in neglect and animal welfare cases are typically those that infest wounds or live on the body surface. The infection of wounds with the larvae of Diptera is known as 'wound myiasis' and usually involves the maggots of blowflies and fleshflies. If untreated, an infested wound would be extended and more flies attracted to lay their eggs (Figure 6.15). Wound myiasis caused by fleshflies can develop very quickly because the female flies deposit first instar larvae within and beside the wound. In addition to causing pain and distress, the condition may give rise to septicaemia (infection of the blood with bacteria) that is potentially fatal. Anybody with an open wound who is incapable of looking after themselves, such as the very young, very old, very ill and those who are mentally incapacitated, is at risk of suffering from myiasis and the discovery of maggot infestation would usually be considered a sign of neglect. Wound myiasis is also a common problem amongst farm animals such as sheep and goats

Figure 6.15 Severe blowfly-strike in a sheep. The infection began as a breech strike and is extending to include the upper hind-quarters

and pets such as rabbits, but persons are seldom convicted under animal welfare legislation unless it has been allowed to develop extensively.

Pharaoh's ants, *Monomorium pharaonis*, and some other ant species will invade wounds and dressings and large numbers can arrive in a short period of time. They are extremely small and it is possible for a large colony to establish itself in a building (they do not live in the wild in the UK) before it is noticed. Pharaoh's ants are capable of transmitting a range of pathogenic bacteria, including various species of *Staphylococcus, Clostridium, Pseudomonas* and *Salmonella*, and therefore their presence in hospitals is a serious matter.

Soiled clothing or dressings, unless changed regularly, will attract the attentions of flies, including non-invasive species such as *Musca domestica* and *Fannia canicularis* and those such as *Calliphora vicina* that are capable of subsequently invading underlying tissues and causing wound myiasis. Fly eggs are usually white or pale yellow and are easily overlooked when seen against a similarly coloured background, such as a nappy or bandage. The eggs hatch quickly owing to the warmth of the body and the larvae may be well developed before the infestation is noticed. Other invertebrate species may also feed on soiled clothing or dressings but they are less useful for determining the length of time the person has been neglected. For example, cockroaches retire to resting sites in crevices, etc. after feeding.

Fleas, lice and mites are common ectoparasites (parasites that dwell on the body surface) of humans living in unsanitary conditions and may sometimes be a sign of neglect. However, an exception is the human head louse, *Pediculus humanus capitis*, which is common even amongst affluent, healthy families. Most domestic and wild animals are infested with ectoparasites although, provided that the animal is healthy, their numbers remain low and are not harmful. Sheep scab mite, *Psoroptes ovis*, is an exception and can rapidly develop into potentially fatal infestation. It is a notifiable disease throughout the UK (i.e. it is a legal requirement to report the presence of the infection to the authorities) and sheep dipping to control it was compulsory at certain times of year whether or not the flock was infested. However, nowadays, sheep dipping remains compulsory only in certain regions. It is not an offence to own an infested animal but it would be an offence not to report its presence or to treat the infestation having discovered it. Many factors influence the progression of sub-clinical sheep scab infestations, which are extremely difficult to detect, into the debilitating active phase – which may occur anything from a few days to over a year later. Consequently, it can be difficult to prove longstanding neglect because the sheep with the most extensive lesions may not be the ones with the oldest infestations.

The role of invertebrates in food spoilage and hygiene litigation

As societies develop, people become less and less willing to accept the presence of invertebrates in their food or the evidence that they have been present in the

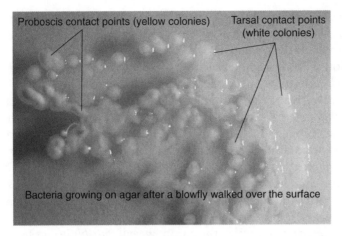

Figure 6.16 A blowfly walked across this agar plate and bacteria have grown where the tarsi and proboscis came into contact with the surface

past. Consequently, even small numbers of invertebrates, their faeces, cast skins or bite marks can be sufficient to render a foodstuff unsuitable for human consumption. This can lead to the loss of large sums of money by the food retailer owing to an inability to sell the product or from legal proceedings resulting from being prosecuted for selling goods unfit for purpose. In addition to the problems caused by the physical presence of invertebrates, many of them are considered harmful owing to their ability to transmit diseases. For example, when insects walk over the food they can transfer large numbers of pathogens such as bacteria, viruses, protozoa and nematodes. Numerous species of invertebrates are associated with food spoilage and only a few will be mentioned here.

Flies, and in particular blowflies, are notorious for spoiling meat and meat products and thereby rendering them unsuitable for human consumption. Spoilage may be caused by adult flies, their eggs and by the larvae. Adult blowflies commonly regurgitate their gut contents whilst feeding and defaecate close to or on their food. Because adult blowflies also feed on faeces and rotting organic matter, they can passively transmit a wide range of human bacterial and protozoan pathogens (Figure 6.16). Many pathogens pass unharmed through the flies' gut or are transmitted on the flies' feet. The stains produced by fly faeces are easy to spot on a pale background but easily overlooked on dark material such as fresh meat. Blowflies often leave large egg masses containing over 100 eggs on meat, usually in damp crevices and under-surfaces rather than exposed regions. The meat is then said to be 'fly-blown' and is unfit for human consumption. Meat that is left exposed for even a short period may become fly-blown if measures are not taken to keep flies out of the shop or kitchen. Unless refrigerated, blowfly eggs usually hatch within 24–48 hours and the larvae develop rapidly. The larvae tend to feed close together and avoid the light so they may not be seen until after the meat is purchased and legal cases then arise if there is a dispute about when the meat became contaminated. These legal cases can usually be resolved by correct identification and knowledge of the conditions under which the meat was kept from the time the animal was killed to the stage at which the dispute arose.

In addition to blowfly and fleshfly larvae, the maggots of various other fly species may be found in a wide variety of food products. For example, the larvae of *Piophila casei*, which are often associated with corpses at the late stages of decay, are also a pest of cheeses and hams. The common name of the larvae is the 'cheese skipper' from their ability to spring over 20 cm into the air. A number of Diptera species are pests of agricultural crops and their larvae may be found on a variety of crops (e.g. the onion fly *Delia antiqua*, and the cabbage root fly *Erioischia brassicae*). Their presence is usually far less serious than blowflies on meat and the affected parts can be cut away so legal cases are less likely to arise.

Many beetles, for example, the grain weevil *Sitophilus granarius* and the mealworm *Tenebrio molitor*, are common pests of stored products. Large populations can build up leading to damage to the grain, etc. and the sacking it is stored in. The problems are becoming worse owing to increasing insecticide resistance among strains of various coleopteran stored product pests. Legal claims arise over disputes at what point a batch of stored product became infested with the beetles.

Ants, bees and wasps are often attracted to sweet substances, such as sugary drinks and foods, whilst social wasps and some ant species are also attracted to meat products. The presence of ants, bees and wasps in food may therefore be a consequence of contamination before or after processing. Parasitoid wasps are often found in stored products as a consequence of their parasitizing pests such as beetles and lepidopteran larvae. Some parasitoid wasps have extremely long ovipositors that may lead to concern over the possibility of being stung – however, this is usually not a real risk.

A number of lepidopteran species have larvae that are pests of agricultural crops (e.g. the cabbage white butterly, *Pieris brassicae*) or stored products (e.g. the flour moth, *Ephestia kuhniella*). The presence of caterpillars on fresh produce is seldom a serious concern because they are easily seen and, along with their faeces, they can be washed off but the presence of stored product pests may give rise to disputes about when the infestation developed (Figure 6.17).

Most cockroach species are large insects but because they are nocturnal evidence of their feeding activity is more obvious than the insects themselves. They are omnivorous and will eat everything from meat products to fruit, wax, paper and grain. Through contamination, they are capable of spreading a large number of human bacterial and protozoan pathogens, so evidence of their presence is a concern for public hygiene. Parts of cockroaches (e.g. legs, wings, egg cases) may occur in processed foods owing to the food being infested during storage or the processing routine. As with all invertebrates, food processing will affect the appearance of the insect body parts and therefore whether they were mixed before or afterwards, although more work needs doing in this area.

Figure 6.17 Typical damage to stored products caused by lepidopteran caterpillars

Psocoptera are very small insects and, although common in houses, their size means that they are often overlooked until their populations reach high levels. They often feed on fungi growing on foodstuffs stored in damp conditions. Their presence in large numbers is therefore an indication that the food has been incorrectly stored. In addition to the damage they cause, Psocoptera have been implicated in causing allergic reactions such as rhinitis and bronchial asthma. Silverfish (*Lepisma saccharina*) and firebrats (*Thermobia domestica*) (both order Thysanura) are also common household pests, although their nocturnal habits means that householders may not be aware of their presence. They are typically found in kitchens, especially in cupboards and around the cooker. They do relatively little damage but occasionally one gets trapped in a bowl and then, without the cook noticing its presence, a tin of food is opened and emptied on top of it. The cook then believes that the tin of food was contaminated. This can be disproved if the insect is still alive or from its state of preservation.

There are numerous mite pest species but because of their small size they are often overlooked. Mites, such as the grain mite *Acarus siro*, can be a serious problem when present in stored products such as flour. The mite's faeces and cast skins cause flour, grain, etc. to develop a musty taste and can cause digestive upset. People routinely handling contaminated foodstuffs can develop allergic reactions resulting in eczema and asthma. Infestations develop when the grain, etc. is stored incorrectly, especially if it is allowed to become damp. Legal cases arise when there are disputes about when (e.g. processing, storage, transport) the food product became infested. This can be determined by estimating the mite population and then, from a knowledge of the biology of the mite and reference to the environmental conditions under which the product had been stored, it would be possible to estimate the time taken to reach this level. Legal cases may also arise when workers claim to have suffered ill health through being exposed to infested grain. However, this would be a medical question and therefore beyond the remit of a forensic entomologist because many other substances (e.g. fungal spores) could cause allergic reactions.

Spiders are all carnivorous and do not harm the food they are found in. However, large tropical spiders, such as tarantulas, are occasionally imported with exotic fruit and are capable of inflicting painful bites if disturbed and their setae (body hairs) are barbed and easily break off and can cause a severe rash if they penetrate the skin. Certain species of spider are prone to falling into food-stuffs that are left exposed to the air, e.g. Wolf spiders, which do not use webs and actively search for their prey, and male house spiders (*Tegenaria* sp.) that emerge from crevices at night to search for females or water. Consequently, spiders are more likely to find their way into food after it has been purchased and placed in the kitchen.

There are a number of common pest species of slugs and snails and they can be found on a wide variety of crops (e.g. potatoes, lettuce, strawberries). They are easy to spot and do not present a serious health hazard should they be consumed accidentally. Small numbers of slugs and snails found on fresh produce bought on a market stall would therefore not be a reasonable ground for complaint.

Invertebrates as a cause of nuisance

When conditions are favourable, certain invertebrates may occur in huge numbers and if they invade homes and offices they can cause considerable nuisance and distress (Figure 6.18). This is sometimes referred to as an outbreak. Favourable conditions may result from natural phenomena, such as the weather, providing ideal breeding conditions or the sudden concentration of insects in an area by convergent wind systems following their migration. These natural outbreaks are seldom long lasting. By contrast, man–made phenomena such as waste-dumps, farms, sewage works, slaughterhouses, etc. provide ideal breeding grounds for some insects and can give rise to more prolonged outbreaks. For example, moth flies (Psychodidae) often breed in large numbers in sewage works and drains and, although they are weak fliers, may be carried by the wind into nearby houses.

Householders living near to farms, waste-dumps or slaughterhouses often complain about large numbers of nuisance insects invading their property and attempts at legal prosecution are sometimes made. However, the insects responsible are usually common species and it is therefore difficult to prove that they originated from the contested site. A mark–release–recapture experiment would be needed to prove that significant numbers of insects were finding their way from the site to the complainant's property and that the complainant was undertaking reasonable precautions to keep them out. For example, it is unreasonable to complain about insects in the house if doors and windows are left routinely open.

Figure 6.18   Thousands of midges congregated in this bathroom overnight, having flown in from a nearby lake

Invertebrates as a cause of structural damage

To wood-dwelling invertebrates, timber that is used in houses or to make furniture, etc. is no different to the fallen logs, etc. in which they would normally live. The consequences of their activity may seriously weaken the strength of the timber, leading to the house becoming structurally unsafe or the furniture collapsing. The consequences of this may lead to large financial losses or accidents in which persons can be seriously hurt.

There are a large number of species of Coleoptera that live in wood; usually it is the larvae that do the most damage as they tunnel through the timber. Damp wood, especially that already affected by fungal decay, is most vulnerable. Sometimes there is no indication that the wood is infested until the adult beetle emerges. Some species are long-lived (3–15 years) and the adult beetle may burrow out long after the tree it was living in was felled, cut up, transported to another country and fashioned into furniture. In the absence of the insects, it is sometimes possible to identify the beetles causing damage to wood from the bore dust deposited around the holes (Figure 6.19). For example, the death-watch

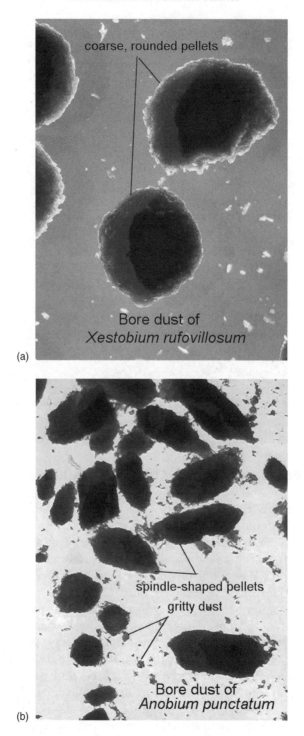

Figure 6.19 Bore dust produced by the larvae of the beetles *Xestobium rufovillosum* (death-watch beetle) (a) and *Anobium punctatum* (furniture beetle) (b)

beetle *Xestobium rufovillosum* produces coarse, rounded pellets, the powder post beetle *Lyctus brunneus* produces a soft fine powder and the furniture beetle *Anobium punctatum* produces spindle-shaped pellets mixed with gritty dust.

Larvae of the wood wasps (*Siricidae*) develop in fallen logs and may also be found in structural timber but they are seldom in sufficient numbers to cause serious structural damage. However, when the adult wasps emerge, their large size and fearsome ovipositor can induce alarm. Similarly, the larvae of several moth species such as the goat moth (*Cossus cossus*), so called because of the strong smell produced by the larva, and also certain members of the Sessidae (clear-wings) (e.g. lunar hornet moth, *Sphecia bembeciformis*) can cause damage to structural timber but although they are both impressively large insects they are seldom in sufficient numbers to cause more than a few unsightly holes.

Termites are commonly called 'white ants' but they are only very distantly related to them – they are much closer to the Blattaria (the cockroaches). All termites are social insects, although the colony size varies between species and may be less than 100 individuals or over 1 million. They are not native to the UK, although there are reports of a colony establishing itself in the south of England. Unlike ants, in which the non-reproductive castes are all female, in termites they consist of both males and females. Termites feed on wood, vegetation, fungi, humus, etc. Some termites prepare fungal gardens in their nests. Termites are a major problem in Africa, Asia and the warmer regions of America. They can rapidly destroy buildings. The ways in which they attack wooden structures varies between species, but usually they hollow out an object leaving just the outer shell. Sometimes the wood is replaced by soil.

For a summary of all the major insect orders with forensic associations, see Table 6.2.

Table 6.2 Summary of major insect orders with forensic associations

Insect Order	Forensic association
Diplura	Found on corpses
Collembola (springtails)	Found on corpses
Isoptera (termites)	Food spoilage, structural damage
Blattaria (cockroaches)	Found on corpses left indoors, food spoilage, nuisance
Psocoptera (plant lice)	Food spoilage
Anoplura (lice)	Neglect
Siphonaptera (fleas)	Neglect
Trichoptera (caddis flies)	Larvae found on corpses left in lakes, ponds or streams
Lepidoptera (butterflies and moths)	Found on corpses, food spoilage, structural damage, illegal trade in protected species
Diptera (flies)	Found on corpses, neglect, food spoilage, nuisance
Hymenoptera (bees, ants, wasps)	Found on corpses, neglect, food spoilage, cause of death, structural damage
Coleoptera (beetles)	Found on corpses, food spoilage, structural damage

Illegal trade in protected species of invertebrates

Large and colourful invertebrates are frequently acquired by collectors or sold for decoration. Some specimens change hands for large sums of money and there is a worldwide market that, for the most part, is perfectly legitimate. Unfortunately, owing to habitat loss and, sometimes, excessive collecting from the wild, the numbers of some species are declining rapidly. This increases the 'value' of collectable species and the illegal sale of plants and animals is estimated to be worth billions of pounds per year on a worldwide basis. Some collectors are not content with obtaining one or two representative examples of a species and will amass as many specimens as they can – this can cause further serious damage to small local populations. Consequently, an increasing number of invertebrates are achieving protected status that makes it an offence to collect, trade or import them. The maximum prison sentence for illegal trading in wildlife has recently increased from 2 to 5 years in the UK. In the UK, the Wildlife and Countryside Act, 1981, names 14 species of insect and 13 species of other invertebrate. Under this Act, it is illegal to capture, kill or sell them, except under license. Possession of any of these species, alive or dead, is considered illegal unless they were acquired before the Act came into force or under license. Examples include the mole cricket (*Gryllotalpa gryllotalpa*), the Large Blue butterfly (*Maculinea arion*), the swallowtail butterfly (*Papilio machaon*), the medicinal leech (*Hirudo medicinalis*) and apus (*Triops cancriformis*) (Figure 6.20).

Quick quiz

1. State four sites where a female blowfly would typically lay its eggs on a corpse and explain why these sites would be chosen.

2. How would you know when a final instar blowfly maggot had finished feeding?

3. How do *Calliphora vicina* and *Protophormia terraenovae* differ in their choice of pupation site? Why is this of relevance in the collection of forensic evidence?

4. Explain why although adult female blowflies are often seen resting and feeding on a corpse in the later stages of decomposition they do not lay their eggs upon it.

5. Explain why a buried corpse may contain thousands of maggots of *Conicera tibialis* but none of *Lucilia sericata*.

6. At what stage of corpse decomposition do Dermestid beetles tend to appear?

7. Explain how ants can affect the rate of corpse decomposition.

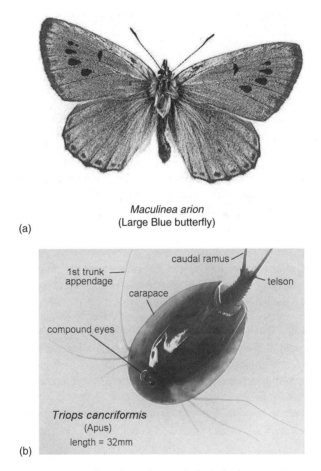

Maculinea arion
(Large Blue butterfly)

(a)

Figure 6.20 Protected species of UK invertbrates include both attractive species such as the Large Blue butterfly (*Maculinea arion*) (a) and the less appealing Apus (*Triops cancriformis*) (b)

8. What would the sight of numerous live body lice moving across the clothing of a dead person indicate? Explain your reasoning.

9. Medical evidence indicates that a murder victim has been dead for only 24 hours but the blowfly larvae recovered from the body were estimated to be about 44 hours old. Is there a conflict of evidence? Explain your reasoning.

10. Briefly distinguish between how a wasp may be either a direct or an indirect cause of human fatalities.

Project work

Title

The effect of burning on invertebrate colonization of a dead body.

Rationale

Murderers often attempt to dispose of a dead body by burning. They are seldom successful but it does alter the chemical and physical nature of the body and this may impact upon the speed of colonization and the rate at which decomposition occurs. This in turn may require allowances when calculating of the minimum time since death on the basis of blowfly growth characteristics.

Method

The problem may be investigated in the laboratory and the field. In the laboratory, blowfly maggots may be reared on meat that has been subject to varying degrees of burning. The growth characteristics (e.g. length, instar) of the maggots should be monitored at least once a day along with the condition of the meat (e.g. change in weight). Blowfly adults may also be offered choices of burnt and unburnt meat on which to oviposit – this could also be tested using the classical Y-shaped choice chamber apparatus. The experiment could be replicated in the field, in which case the environmental conditions should also be recorded and also the invertebrate species composition. The experiment could be repeated at different times of year and also the meat might be covered in clothing material before it was burnt to determine whether this had any impact.

Title

The effect of competition on the growth characteristics of blowfly maggots.

Rationale

Blowfly maggots are usually in competition with one another and this may result in either an increase or a decrease in their growth characteristics. This in turn may require allowances when calculating the minimum time since death on the basis of blowfly growth characteristics.

Method

Single and mixed species of maggots are allowed to develop on a fixed amount of food. The amount of food will be such as to ensure that the highest densities of maggots are at risk of consuming all the food before the end of their final instar. The numbers of maggots would be assessed daily to determine if and when cannibalism took place and the number of maggots that successfully pupate and emerge as adults. The length and instar of the maggots would be assessed at least once a day.

7 Forensic information gained from invertebrates

Chapter outline

The importance of correct identification

Calculating the minimum time since death or infestation from invertebrate development rates

Complicating factors affecting minimum time since death calculations

Determination of the minimum time since death or infestation using invertebrate species composition

Determination of the minimum time since death using ectoparasites

Determination of body movement or point of infestation from invertebrate evidence

Invertebrate evidence in cases of wound myiasis and neglect

Detection of drugs, toxins and other chemicals in invertebrates

Obtaining DNA evidence from invertebrates

Determining the source and duration of invertebrate infestations of food and timber products

Objectives

Explain the vital importance of correct species identification for the validity of forensic evidence based upon the presence of invertebrates.

Describe how the minimum time since death or infestation may be calculated on the basis of the species of invertebrates present and their stage(s) of development.

Discuss the limitations of minimum time since death / infestation calculations, the factors that can influence them and how potential problems can be overcome.

Describe how invertebrates can be used to determine whether a body or object has been moved and to provide pharmacological and molecular evidence.

Explain how it is possible to determine where and when invertebrate infestations of food and timber products took place.

Essential Forensic Biology, Alan Gunn
© 2006 John Wiley & Sons, Ltd

The importance of correct identification

Owing to their abundance and diversity of lifestyles, invertebrates can provide a wide variety of forensic evidence ranging from estimating how long a person has been dead to determining whether the owner of burger shop is guilty of failing to comply with local hygiene regulations. However, the common factor is that the quality of the evidence depends upon accurate identification and a thorough understanding of invertebrate biology. For example, if the species of blowfly involved in a murder case is incorrectly identified it could result in the estimated time since death being either too short or too long, thereby conflicting with other evidence. Similarly, if the insects found in a foodstuff are misidentified, it could result in the blame for the infestation being attached to the wrong party in the supply chain. Furthermore, if one does not understand the biology of the organisms that are to be used as evidence, then one may not collect them in an appropriate manner or come to false assumptions.

Calculating the minimum time since death or infestation from invertebrate development rates

If one has reliable information on the biology and development times of an invertebrate and how these are affected by temperature and other environmental conditions it is possible to back-calculate to determine when the egg from which a specimen developed was laid. This can be useful in determining when a person died, a wound myiasis developed, food became contaminated or a piece of furniture became infested. The technique is heavily dependent upon correct identification, the accuracy of the estimated environmental conditions the invertebrates are thought to have experienced and the reliability of the experimental development data used to perform the calculations. It is called the 'minimum' time because the estimate reflects the least time it would take for the invertebrate to reach a particular stage of development — it is possible that the person might have been dead for longer or the food was infested even earlier but it is very unlikely that the initial infestation took place after the calculated date. In some texts the time is referred to as the 'earliest oviposition date' (EOD), which is more accurate and reflects the fact that a dead body does not have to be involved.

Blowflies are the most commonly used invertebrates to determine 'minimum time since death' calculations but the procedure would be the same for any other creature. Maggots are collected from all sites in, on and around the body, ensuring that these include the oldest (largest) maggots and any pupae or adults that are also present. During the early stages of decomposition, the most mature maggots or pupae indicate the minimum time since death because they will have developed from the eggs (or larvae in the case of fleshflies) deposited on the body immediately after the person died. The morphology of the posterior spiracles and the cephaloskeleton changes between instars and therefore provides an indication of age (Figure 7.1). An estimation of age can also be made from measuring the length and / or dry weight of the larvae. Both length and dry weight increase with age until the larvae cease feeding and wander off in search of a pupation site — at which point length and weight decline. Some workers also measure the length of the crop in the final instar because this shrinks when feeding ceases. Although the larvae provide the most accurate basis for estimations of minimum times since death, it is possible to age the pupae, especially those of cyclorrhaphan flies (e.g. blowflies, fleshflies and house-flies), if pupation has commenced when the body is discovered. When the maggot moults to the pupal stage, the third larval instar cuticle is retained and serves as a protective puparium. During this process the insect contracts and the puparium

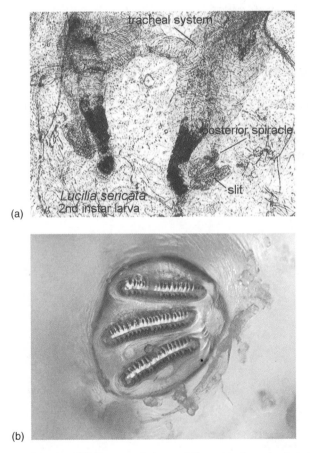

(a)

(b)

Figure 7.1 Compare the very different morphology of the posterior spiracles of the second (a) and third (b) instars of the blowfly *Lucilia sericata*. In the second instar, blowfly larvae have two slits in their posterior spiracles whilst in the third instar there are three slits present

darkens as it becomes tanned and sclerotinized over a period of several hours. The extent of the tanning is therefore an indication of age (Figure 7.2). Within the puparium, the insect first enters the pre-pupal stage before metamorphosing to the pupal stage — this can be demonstrated by dissecting the pupa from its puparium. The possibility of determining the age of a pupa by assessing the activity of temporally expressed genes (i.e. those that are switched on or off at specific points in development) is being investigated but it is too soon to be able to say how applicable the technique will prove. It is possible to distinguish newly emerged cyclorrhaphan flies from those that are older and, albeit crudely, how long they have been emerged. This may be useful where a body at an advanced stage of decay is discovered. Newly emerged flies are pale in colour, their wings may not be fully expanded and the ptilinum (a sack-like structure on the head) may either be expanded or not yet fully retracted behind the ptilinal suture (Figure 7.3). This state lasts for only a few hours. It is difficult to determine the age of male flies but in females the ovaries are immature at emergence and the eggs develop over a period of days. Wings tend to become abraded with age and have been used as an age estimate in ecological studies (Hayes *et al.*, 1998). The levels of pteridine eye pigments are known to alter with age in a variety of fly species, including *Lucilia sericata*, but the technique is not used commonly in forensic studies. The different techniques that can be used to age adult flies have been reviewed by Hayes and Wall (1999), whilst Smith (1986) provides an excellent description of blowfly larval stages and a series of forensic case studies. Table 7.1 summarizes the age indicators in cyclorrhaphan flies at different stages of development.

Once the lengths and the instars of the maggots are known, it is possible to determine how long it took to reach that stage under controlled conditions using

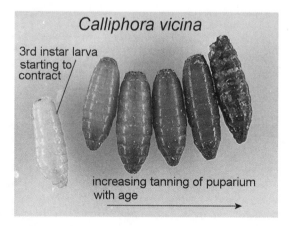

Figure 7.2 Blowfly pupae of varying ages. Note how the cuticle of the final instar larva is retained to form a puparium and becomes increasingly tanned and sclerotinized. The degree of darkening can provide an indication of age

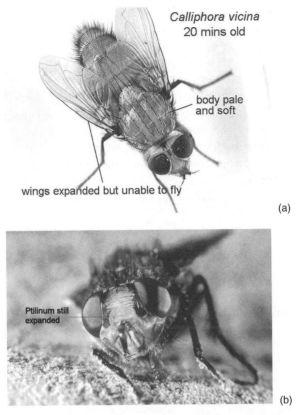

(a)

(b)

Figure 7.3 Characteristic features of a recently emerged blowfly. The body is pale and even if the wings are fully expanded it is unable to fly (a) and the ptilinum may not be fully retracted (b)

Table 7.1 Summary of age indicators in cyclorrhaphan flies at different stages of development

Life cycle stage	Age indicator
Larva	Body length Crop length (post-feeding stage only) Shape of cephaloskeleton Shape of posterior spiracles Shape of anterior spiracles
Pupa	Coloration of puparium (initial stages of pupation only) Stage of development of pupa within puparium
Adult	Ptilinum (expanded or retracted) Wings (not expanded in recently emerged, increasingly abraded with age) Hardening and darkening of cuticle immediately after emergence Development of ovaries Pteridine pigment accumulation in eyes

published studies on rates of development and estimates of the conditions likely to have been experienced by the larvae (see Chapter 6). Correct identification is essential because even closely related species may develop differently under the same conditions. The time period taken to reach the maggot developmental stages found on the body must then be adjusted to fit the site where the body was found. One way of doing this is to convert the temperatures and times into accumulated degree hours (ADH) or accumulated degree days (ADD) by multiplying the time by the temperature (°C). Because the time required for development increases as temperature decreases, the total number of ADH required to reach any given stage remains constant. For example, according to Greenberg and Kunich (2002), at 25°C the average minimum durations of the developmental stages of *Calliphora vicina* are as follows:

Egg stage = 14.4 hours
First instar = 9.6 hours
Second instar = 24 hours
Third instar = 158.4 hours

To calculate ADH, multiply the sum time taken to reach a particular stage by the temperature. For example, the ADH taken to reach the third instar is:

$$(14.4 + 9.6 + 24) \times 25 = 1200 \text{ ADH}$$

To convert ADH to ADD divide by 24h. For example, the ADD taken to reach the third instar is:

$$1200/24 = 50 \text{ ADD}$$

1. Suppose that a body is discovered at 11.00 a.m. on 23 June and that insect specimens are collected at midday on the same day.

2. The most mature *Calliphora vicina* larvae found on the body are just moulting from the second instar to the third instar.

3. The ADH is the total ADH required to complete the egg stage and all of the first and second instars:

$$(14.4 + 9.6 + 24) \times 25 = 1200 \text{ ADH}$$

4. To estimate the period of activity, work backwards from the time the maggots were collected (12.00 a.m. on 23 June). There were 12 hours of development between midnight and 12.00 a.m. on the day. The mean temperature at the scene for that set period was recorded as 15°C. This means that a total of 180 ADH were accumulated on 23 June (12 hours × 15°C).

5. On 22 June, the mean temperature was 12°C, and this gives a total ADH of 288 (24 hours × 12 = 288 ADH).

6. On 21 June, the mean temperature was 13°C, and this gives a total ADH of 312 (24 hours × 13 = 312 ADH).

7. On 20 June, the mean temperature was 13.5°C, and this gives a total ADH of 324 (24 hours × 13.5 = 324 ADH).

8. Add the totals: 180 + 288 + 312 + 324 = 1104 ADH.

9. 1200 − 1104 = 96 ADH still outstanding.

10. On 19 June, the mean temperature was 13°C, so each hour accounted for 13 ADH.

11. Divide 96 by 13 = 7.4 hours of development on 19 June.

12. Counting backwards from midnight, this means that gravid (those ready to lay) female flies must have laid their first eggs between 4 and 5 p.m. on 19 June.

It is not always necessary to calculate separate ADHs for every individual hour of the day because the temperature of a whole corpse does not change as rapidly as air temperature. However, allowances may need to be made if the body has been dismembered and only small body parts are found. Some workers have attempted to improve the accuracy of their estimates by incorporating computer models of insect growth into their calculations (Byrd and Allen, 2001).

Another way of determining the age of a maggot is by comparing its length with data in an isomegalendiagram (Grassberger and Reiter, 2001). An isomegalendiagram is obtained by plotting temperature (y axis) against time taken for a maggot to reach a given length (x axis). It therefore consists of a series of curves from which one can read off, for example, that a maggot measuring 5 mm might have taken 60 hours to reach that size at 16°C but only 24 hours at 25°C. Although an extremely neat method, generating the data to prepare an isomegalendiagram would be very time consuming.

It is important to remember that even with the most rigorous collection and analysis of evidence any biological process usually involves a great deal of 'noise' as a consequence of various complicating factors. Estimates of the minimum time since death are precisely that — estimates — and the amount of variance around an estimate will increase with the length of time that a person has been dead (Archer, 2003; Nowak, 2004).

Complicating factors affecting minimum time since death calculations (Table 7.2)

Table 7.2   Summary of factors complicating the minium time since death calculations from blowfly larval development rates (EOD = earliest oviposition date)

Factor	Reason
Restricted access of adult flies to body	Delay in egg-laying so EOD estimated long after death
Maggot feeding ball present	Elevated local temperature means that maggots develop faster than expected from environmental temperature so EOD estimated before death occurred
Many Diptera species present	Different species develop at different rates. Biggest may not be oldest, it may simply grow faster
Food quality	Low food quality may delay larval development
Pupal diapause	Pupal development prolonged
Pupal disease / parasitoids	Adults may not emerge
Predators and competitors	Larval population has difficulty establishing itself. EOD estimated long after death
Drugs and toxins	May speed up or slow down larval development
Chemical fly deterrents	Delays egg-laying so EOD estimated long after death
Myiasis	EOD estimated before death
Body at advanced decay	No maggots or second generation present
Low environmental temperature	Uncertainty at what temperature larval development ceases
Microclimate effects	EOD estimated too short or too long in relation to time of death owing to using incorrect development temperature
Incorrect storage	Maggots shrink and therefore appear younger than they really are so EOD estimated long after death
Laboratory development rate data unreliable	Larval development rate may be estimated incorrectly so EOD estimated too short or too long in relation to time of death

1.   The body is discovered shortly after death and although blowflies have begun laying eggs these have not yet hatched.

Solution

The eggs can be collected and stored in a controlled environment until hatching. Bourel *et al.* (2003) have demonstrated that for the blowfly *Lucilia sericata* it is possible to estimate the time of egg-laying with an accuracy of about 2 hours by relating the time taken for the eggs to hatch to the environmental conditions they experience. It would be interesting to know whether this approach would be suitable for other fly species. It would be remembered that sarcophagid flies 'lay' larvae rather than eggs so it cannot be assumed that where both eggs and maggots are found together they belong to the same species. This approach would be particularly appropriate to disputes concerning neglect (e.g. time since a would was dressed) or food hygiene cases in which blowfly egg are likely to be the principle evidence.

2.   Blowflies may not have had access to the corpse immediately after death owing to the way it was stored before disposal. For example, it may have been kept in a sealed container or the boot of a car.

Solution

Consider that estimates of minimum time since death based on blowfly larvae data may be too short. Furthermore, if the body was already at an advanced state of decay when it finally became exposed, blowflies may not have been attracted to it.

3.   During putrefaction, large numbers of maggots collect together to form a feeding mass (sometimes called a 'maggot feeding ball'), which, together with microbial action, generates remarkable amounts of heat (Figure 7.4). This can raise the temperature in the centre of the feeding mass by more than 20°C above ambient, the extent tending to depend upon the size of the dead body and the number of maggots. Large bodies can support high maggot population densities and thereby produce a temperature excess above ambient within the feeding mass. Consequently, a large body might decompose at a faster rate than one of smaller size.

Solution

There is some dispute in the literature about the impact of a large feeding ball on maggot development times and whether allowances should be made and, if so, how. Some workers consider accumulated degree hours (ADH) calculations to be reliable because it takes several days for a large feeding mass to develop. Furthermore, it is only at the centre of the maggot mass that the temperatures are very high and individual maggots are not thought to spend long there because it is potentially harmful — although proving this by tracking the

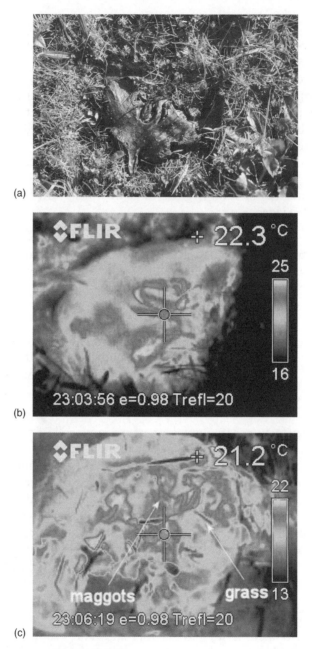

Figure 7.4 (a) Decaying liver left on the surface of the soil. (b) Thermal camera image of the upper surface of the decaying liver. (c) Thermal camera image of the lower surface of the decaying liver. Note the localized 'hot spot' regions on both the upper and lower surfaces where the temperature is considerably higher than that of the surroundings

movements of individual maggots within a feeding ball would present an interesting challenge! Anyway, it is a good idea to be prepared for criticism and if the largest maggots are those recovered from within a feeding ball and a record has been made of the temperature within the ball then it would be sensible to calculate two minimum development times: one incorporating several hours of development within the ball (determined on the basis of the ball's size and hence how long it has held an elevated temperature) and one without.

4. Many species of maggot are usually present on a dead body.

Solution

It is sensible to determine minimum periods since death separately for several species and from larvae collected from different sites on the body to gain an accurate assessment. This is particularly important in countries with strong seasonal climates, such as the UK. For example, Erzinclioglu (2000) has described a case in which the development stage of blowfly larvae found on a corpse, discovered in the spring, suggested a minimum period since death of about 2 weeks. However, the presence of large numbers of the larvae of winter gnats (Diptera, Trichoceridae) indicated that the person had died much earlier, during the winter.

5. The availability of food and its chemical composition can affect the size of the maggots and their development times (Green *et al.*, 2003). Blowfly larvae fed a restricted diet may take longer to develop whilst third instar larvae that have their food removed may pupate at a smaller size than they would normally.

Solution

A fresh, whole, human body will present maggots with more than enough food for at least the first generation of maggots to complete their development. However, separated body parts, such as hands, may present a more limited food resource, especially if the maggot (etc.) population is high. The possibility that food restriction may be affecting the developmental rate or size of the maggots needs to be considered.

6. The pupae of some fly species may enter into a diapausing state (a form of dormancy) in response to conditions they experienced earlier in their development. The development of these pupae is therefore prolonged and the adult flies may not emerge for weeks or months.

Solution

A sample of pupae should be dissected to determine whether they are alive but in diapause or death has occurred.

7. Owing to disease, parasitoids, etc. many pupae may die before the adult flies emerge.

Solution

It is important to collect large sample sizes and to check that the pupae are actually alive before using them to estimate the minimum development time.

8. Predators are often attracted to feed on the detritivores found on corpses and can slow the decomposition process by carrying off the eggs and larvae of maggots, as well as killing adult flies arriving to lay their eggs (Archer and Elgar, 2003). In addition, some fly maggots feed on competitors within a dead body or a wound of a living animal. For example, the larvae of *Chrysomya albiceps* are well known for their predatory behaviour. This species is not present in the UK but it is found in central Europe and its distribution may be moving northwards and this could impact upon forensic investigations (Grassberger *et al.*, 2003).

Solution

It is necessary to collect all the invertebrates found in, on and around the body and to estimate their abundance. One can then make judgements on whether the predators were likely to have an impact.

9. The effects of drugs and toxins on the rates of development of blowfly larvae are poorly understood. Studies have indicated effects that have ranged from zero to either speeding up or retarding the rate of development (e.g. Bourel *et al.*, 1999). For example, maggots of the fleshfly *Boettcherisca peregrina* feeding on tissues containing morphine develop faster, grow bigger and produce pupae that take longer to develop resulting in an error of up to 29 hours if the estimated postmortem interval is based on larval development and 18–38 hours if based on the duration of the pupal stage (Goff *et al.*, 1991).

Solution

There are numerous drugs and toxins and numerous species of maggots associated with corpses, so it is impossible to generalize on the effects of their interactions. Furthermore, persons who die of drug overdoses may contain a cocktail of different chemicals and the drugs may not be evenly distributed between the tissues of the body. For example, persons whose death is associated with the drug ecstasy are often found to have consumed alcohol and other drugs, such as cocaine, amphetamines and heroin. One can therefore only keep an open mind that chemicals within the body may have influenced maggot development to some extent.

10.   The murderer may attempt to confuse the evidence by spraying the corpse with a chemical that deters blowflies from laying their eggs.

Solution

Chemicals vary in their effectiveness at reducing blowfly egg-laying. For example, sodium hydroxide appears to have little effect whilst petrol or patchouli may delay egg-laying by one or more days (Bourel *et al.*, 2004) but a great deal will depend upon the concentration and the amount of the chemical used. The presence of chemicals may be indicated immediately by their smell, their obvious physical presence (e.g. as a powder or crystalline deposit) or the damage they have caused to clothing and body tissues. They can be tested for by routine chemical analysis and it would then require experimentation to determine whether they might have affected insect colonization.

11.   If maggot activity commenced before the person died, the estimation of minimum time since death may be too long. For example, if a person is suffering from wound myiasis, they may subsequently die of their wounds or from the potentially fatal consequences of wound myiasis.

Solution

Be aware that when there is a mismatch between the medical or other evidence of the minimum time since death and that derived from entomological analysis, this could be the reason. It would, however, provide evidence that the person was wounded some time before they died.

12.   If the body is at an advanced stage of decay there may be few, if any, blowfly maggots present. If any are found, they would not have developed from eggs deposited soon after the person died and therefore could not be used to determine the minimum time since death.

Solution

Attempt to determine the time since death from other invertebrates or from overall invertebrate species community structure, although this method is less accurate because it is strongly affected by time of year, environmental conditions and locality.

13.   Estimations of larval blowfly developmental rates have assumed a constant, linear relationship between the development rate and temperature. These studies suggest that below a certain temperature, e.g. 5°C, development ceases. However, this assumption is no longer valid (Ames and Turner, 2003) and varieties of the same species of blowfly from different countries, or different parts of the same country, are known to exhibit different temperature sensitivities.

Solution

If the temperature has periodically fallen below the supposed minimum for development, it would be sensible to calculate two minimum times since death firstly on the basis that development ceased and secondly that it had continued. Ideally, larvae collected at the time would be placed at different low temperatures to record the effect on their rate of development. However, owing to the cost and logistical difficulties, this may prove impossible.

14.   The microclimate where the blowfly larvae are found may be different from that only a short distance away.

Solution

Be aware of possible complicating factors, e.g. snow may have acted as an insulator, and record the environmental conditions at the exact place the maggots were found.

15.   Maggots that are stored incorrectly will shrink although the amount of shrinkage will vary with the method used and the species and size of the maggot. This can result in their age being underestimated.

Solution

Ideally, maggots should be measured before they are placed in preservative. Otherwise, a reliable means of estimating the percentage shrinkage that has taken place needs to be made by experimentation and allowed for in any calculations. There are known to be interspecific differences in the effect of killing and preservation on maggot length and changes in length may also occur during storage (Adams and Hall, 2003).

16.   There is surprisingly limited published information on the development rates of insects of forensic importance. For example, do insects of the same species develop at the same rate under identical conditions regardless of their geographic origin? Would insects feeding on skeletal muscle develop at a different rate to those developing in the viscera? Data on other inverte-

brates of forensic importance tend to be even more limited than that on blowflies.

Solution

Compiling data on development rates is time consuming and difficult to publish in high-impact journals. It therefore tends to have a low priority in university research departments and the results, if published at all, are spread across a wide range of periodicals. Ideally, there should be an agreed protocol for gathering information on development rates and a central web-based databank in which workers could enter their findings.

Determination of the minimum time since death or infestation using invertebrate species composition

The invertebrate fauna associated with a recently dead body is very different to that associated with one that has become skeletonized. Similarly, the composition is also influenced by the time of year, so the variety of invertebrates associated with a corpse in December will be different from that in August. It has therefore been suggested that one can estimate the age of a corpse from the diversity of the invertebrate fauna found upon it. In practice, the invertebrate faunal succession may vary markedly between individual corpses, and environmental and local factors can heavily influence when and whether certain insect species arrive (Archer, 2003). Consequently, the species composition is not a very robust method for estimating the minimum time since death or infestation.

Determination of the minimum time since death using ectoparasites

Fleas, lice and ticks move away from a dead body when it starts to cool, so their presence would indicate that a person or animal had not been dead for long. Fleas may be revived if submerged for less than 24 hours and may therefore provide evidence of how long a body has been underwater. Lice tend to be more sensitive and die within 12 hours. The skin follicle mite, *Demodex*, is a common parasite of humans and may sometimes cause pathology either directly or through acting as a vector of disease. It is unable to leave a dead body but it is capable of surviving for over 50 hours after death of the host. However, the lack of a clear relationship between survival rate and time since host death limits its usefulness in forensic investigations (Ozdemir *et al.*, 2003). Like the other ectoparasites mentioned above, its potential to act as a vector of disease needs to be borne in mind by persons working with dead bodies.

Determination of body movement or point of infestation from invertebrate evidence (Table 7.3)

Because many invertebrates have a restricted geographical distribution or occupy specific habitats, their presence can be used to determine the past history of a person or object. For example, the discovery of a corpse in Kent containing an insect belonging to an uncommon species only previously recorded in the North of England would suggest that the body had been moved. Similarly, the

Table 7.3 Summary of insect evidence that a body was moved at some point after death

Movement of body	Insect	Evidence
Exposure followed by burial	Blowfly eggs and larvae	>24 hrs exposure
	Histerid beetle larvae	~2–3 days exposure
	Piophilid fly larvae	~7 days exposure
	Dermestid beetle adults and larvae	>14 days exposure
Burial or placed in a sealed container followed by exposure	Blowfly eggs and larvae	Absent or stage of development younger than expected from degree of decomposition
	Phorid fly larvae	Abundant phorid fly larvae and adults but no evidence of other fly species
Inside to outside	Blowfly larvae and pupae	Body mummified or skeletonized but no evidence of previous blowfly larval activity on the body or pupae in the vicinity
	Dermestid beetles and Tineid and pyralid moths	Both present but no evidence of previous colonization by other insect species
Outside to inside	Blowfly larvae and other detritivore species	Presence of species that do not normally enter buildings
	Any 'accidentally associated' species	Woodland, moorland, etc. species accidentally trapped in clothing
Geographical (between different regions or countries)	Blowfly larvae / other detritivore species or any 'accidentally associated' species	Species found outside their normal geographical distribution

presence of parasitic protozoa and helminths within the body can provide an indication of a person's origins or recent travels. For example, finding the eggs of the schistosome *Schistosoma haematobium* in the urine indicates that a person had been living in certain parts of Africa or the Middle East.

The invertebrates found on a body need not be a species associated with decay but could be a predator or one that used the body or its clothing as a refuge, or it could have accidentally become trapped in the clothing of the person whilst they were alive. Similarly, the finding of an insect with a distribution restricted to Northern Europe in a foodstuff processed and packed in Indonesia would indicate that infestation occurred after the goods reached Northern Europe. The presence of insects can also be used to determine the source of drugs such as cannabis. For example, the presence of foreign species can indicate the country of origin of the cannabis and therefore whether it was imported (Crosby and Watt, 1986). This is important because it affects whether or not the person is convicted of drug trafficking.

Murderers often move the body of their victim. However, a dead body, especially that of an adult, is difficult to handle. Consequently, dead bodies tend to be dragged for a short way from a path or roadside or, if transported any distance, this is done in a vehicle, after which the body is dragged a further short distance to its final resting place. Movement from outside to inside a building might be indicated by the presence of typical moorland or woodland species, whilst the finding in summer of a mummified corpse containing dermestid beetles and tineid moth larvae in a deciduous woodland would indicate that the body was initially stored indoors. When a person dies indoors and the doors and windows are sealed to prevent the access of flies, the body may mummify. In this case, dermestid beetles and tineid moths, which are common household pests and may already be present in the room, will start to infest the corpse once it is dry enough. Mummification is highly unlikely to occur in an exposed corpse left outside in the UK because the climate is too damp. Furthermore, if there was no evidence of blowfly activity (e.g. puparia), it would indicate that the body was placed there a long time after death occurred.

Many invertebrates avoid light and are thigmotactic, i.e. they orientate themselves to vertical surfaces. Consequently, two bodies within a room may be colonized, and therefore decomposed, at different rates (Figure 7.5). A body at the side of a room will tend to be colonized before one in the centre, and a body that is in full sunshine may be colonized later than one in a dark corner. The presence of clothes will affect colonization because invertebrates may be able to move underneath the clothes and gain protection from desiccation and UV radiation, however tight belts, etc. can limit movement and lead to them congregating at the point of restriction.

Even shallow burial will prevent blowflies from colonizing a corpse. The exact depth will depend upon the soil type, moisture content, etc. It therefore follows that if a buried body is discovered to contain the eggs, larvae or pupae of blowflies, the chances are high that the body was exposed to air for a period of

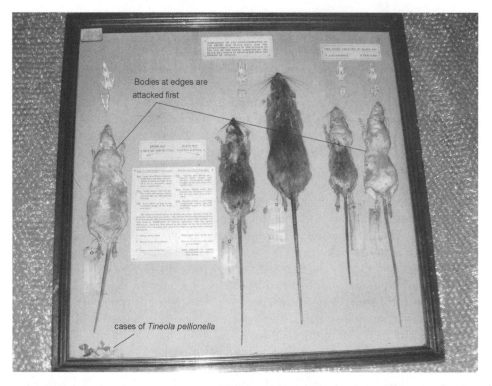

Figure 7.5 These museum specimens illustrate how bodies within the same room or environment may not decompose at the same rate. In this display case, the bodies of the rats closest to the edges of the box were attacked first by clothes moth larvae

time before it was buried. The species composition found on the body can provide an indication of how long the body was left above ground before it was buried. The presence of blowfly larvae would indicate that the body had not been buried for at least 24 hours, histerid beetles 2–3 days, piophilid flies 7 days and dermestid beetles 14 days. Similarly, if a body is found on the surface that contains little evidence of blowfly attack for the time of year and its stage of decomposition, and an invertebrate fauna that would normally be associated with a buried corpse, then it is likely that the body was exhumed.

A corpse (or bits of it) might be moved by a murderer, by the actions of dogs, foxes, badgers, etc. or by natural phenomena such as landslips and rivers overflowing. Invertebrates are not large enough to move a human body but they can bring about changes in the arrangement of any clothes the victim was wearing. Experiments performed on dead pigs of human size dressed in women's clothing have demonstrated that the actions of decay and invertebrate activity could mimic the indications of sexual assault typically observed in homicide cases (Komar and Beattie, 1998). The maggot masses, acting in concert with bloat, moved the clothing: the underwear and tights were moved down the hind legs and the skirt was pushed up.

Invertebrate evidence in cases of wound myiasis and neglect

Determining whether neglect has occurred through a wound, bandages or a nappy becoming infested with maggots requires a similar procedure to that outlined above (Benecke and Lessig, 2001). Neglect cannot be automatically assumed if evidence of insect activity is found because eggs may be laid immediately after routine cleaning and, in the case of fleshflies, because first instar larvae are laid, maggots are instantly present. Larval samples need to be taken, the time noted and a record made of the environmental conditions (temperature, etc.). For example, Gheradi and Constantini (2004) describe a case in which the maggot-infested body of an elderly woman was discovered in the filthy flat she shared with her daughter. This resulted in her daughter being charged initially with 'concealment of a corpse' although she claimed that she had given her mother a meal the previous night. An autopsy later in the day of discovery demonstrated that the body's temperature was 34.7°C, indicating that the woman had not been dead for long, and this was corroborated by finding only the early stages of postmortem decomposition. However, the presence of third instar blowfly larvae and second instar fleshfly larvae on the skin and the diapers indicated that the woman had not been kept clean for several days and the daughter was subsequently charged with 'elderly neglect'.

Detection of drugs, toxins and other chemicals in invertebrates

When a body is discovered and it is believed that drugs or poisons may have contributed to the person's death, there are standard procedures for testing for these substances. However, if the body is not found until the late stages of decay, there may be few or no tissues left for analysis. In these circumstances, it has proved possible to detect the presence of substances in the maggots and their puparial cases (Pounder, 1991; Hedouin et al., 1999; Goff and Lord, 2001). For example, using LC–MS (combined liquid chromatography–mass spectrometry), one can detect nordiazepam residues in a single maggot allowed to feed on a diet containing drug residues equivalent to those that would be found in human skeletal muscle following a fatal overdose (Laloup et al., 2003). However, Tracqui et al. (2004) were unable to relate the concentrations of a wide variety of drugs found in dead bodies with those present in invertebrates feeding upon them. Furthermore, the invertebrates exhibited marked differences in drug residue concentrations between individuals recovered from the same body. Consequently, apart from noting the presence, they concluded that it would be unsafe to draw conclusions from drug residues found in invertebrates recovered from a dead body; the absence of drug residues in the invertebrates should not be considered reliable evidence that there were no residues in the person's body before he or she died.

Roeterdink et al. (2004) demonstrated that blowflies reared on meat containing gunshot residues accumulate lead, barium and antimony although these metals bioaccumulate in the maggots' bodies to different extents. However, more work needs to be done to determine whether gunshot residues would affect the maggots' rate of development and whether forensically useful information could be obtained from the isotopic signature of the metals found within them.

Spiders' webs often trap particles being transported by wind currents. For example, in the Soham murder trial (UK, November 2003) it was reported that dark discoloration of cobwebs was observed on light-fittings close to the bin in which the victims' burnt clothes were found. The coloration of the cobwebs became lighter the further one moved away from the bin. Unfortunately, there does not appear to be a method for testing cobwebs for smoke exposure.

Obtaining DNA evidence from invertebrates

It is possible to extract DNA from the crop of a maggot or beetle and thereby determine what it has been feeding on. This can be useful in cases where there are insects but no body, where there is a suggestion that the insects have crawled onto a body from elsewhere or where the insect samples have been switched (DiZinno et al., 2002; Linville and Wells, 2002 Linville et al., 2004; Zehner et al., 2004a; Campobasso et al., 2005).

Fleas and lice frequently transfer between humans and between humans and animals that come into bodily contact with one another. It is therefore possible to link persons together by analysing the blood meal present in the guts of lice or fleas (Mumcuoglu et al., 2004). For example, crab lice, Pthirus pubis, may be transferred between assailant and victim during a sexual assault and DNA extracted from the blood meal present in their guts can be used to link the two people together (Lord et al., 1998).

Determining the source and duration of invertebrate infestations of food and timber products

After identifying the species found within the food and its stage of development, it is often possible to determine when the infestation initially developed and therefore who was responsible for allowing it to happen. For example, the customer of a UK firm supplying frozen chicken wings to a Caribbean Island was refusing to pay for deliveries because after the wings were cooked they were found to contain 'living maggots'. The claim was considered fraudulent because although some flies will lay eggs on frozen meat (e.g. *Calliphora vicina*), even if the chicken wings had been contaminated by fly eggs during processing the freezing process would have killed them. The transport of frozen foodstuffs is highly regulated and, until the customer removed the foodstuff's protective packaging, no flies should have been able to gain access to the chicken wings. The fact that living maggots were found in the wings after cooking says little for the standard of cooking and also indicates that there was sufficient time for flies to lay their eggs and for these to hatch — a process that would be expected to take several hours even in the Caribbean. The larvae were in their second instar and belonged to a species not present in the UK. Assuming that the chicken wings arrived in a frozen state, they must have been left unprotected in the kitchen. The only other alternative would be that the chicken wings were not transported properly and had become contaminated with eggs between arrival on the island and the kitchen.

When food is packaged, the presence of chewed holes in the goods can indicate whether the insects bored their way into or out of a bag or sack. Furthermore, the presence of only a few holes in the packaging and a population of very young insects plus one or two adults would suggest that the infestation was initiated from outside. The absence of holes would indicate that the infestation began before the goods were packaged and their stage of development could be used to calculate when this took place.

In the case of timber infestations, legal claims for compensation may be brought if the commercial preservation treatment performed on a property failed whilst still under guarantee. The appearance of adult beetles would not necessarily be an indication of an extant infestation because they could have flown or crawled in from elsewhere. The room or furniture would need to be sealed to prevent the entrance of insects from outside, and the subsequent appearance of adults, the presence of bore dust or fresh holes would demonstrate that an infestation was present. Obtaining the larvae from within the wood would enable an estimate of how long the infestation had been present but this is seldom possible because it would cause too much damage. Identification is particularly important if the timber was imported because it can indicate whether infestation took place abroad. It is also necessary to destroy non-endemic species before they are able to establish and breed in the UK.

Quick quiz

1. Why is correct species identification so important in forensic entomology?

2. When collecting maggots from a corpse, why should one be sure to include the largest maggots present?

3. State four morphological features that can be used to estimate the age of a blowfly maggot.

4. How would you distinguish between a newly formed blowfly pupa and one several days older?

5. Given the data listed below, calculate the accumulated degree hours necessary for blowfly species A to reach the second instar.
   Rearing temp = 22°C
   Egg stage = 16 hours
   First instar = 10 hours
   Second instar = 25 hours
   Third instar = 34 hours

6. Briefly explain how the formation of a maggot feeding ball could influence the development rate of blowfly larvae?

7. Briefly describe how the presence of drugs could affect the development of blowfly larvae and why this is relevant to calculations of the minimum time since death.

8. Why should maggot length be calculated before the specimens are placed in preservative?

9. Briefly explain how it is sometimes possible to determine whether food infestation took place before or after the goods were packaged.

10. Briefly explain why species identification can be important in determining the point at which a food product became contaminated with insect body parts.

11. The body lice found on the corpse of a vagrant were still alive when his body was recovered from the river Thames. What does this indicate about the time his body had been in the water? Explain your reasons.

12. Briefly explain how the presence of invertebrates can be used to demonstrate that the body of a murder victim was exposed for several days before it was buried.

13. State two reasons why it might be useful to extract human DNA from blowfly maggots.

Project work

Title

How long can a blowfly maggot live underground?

Rationale

There may be a delay between a victim of crime dying and their body being buried. During this time the body may become infested with maggots but there is little published evidence on the extent to which they are capable of feeding and developing on the corpse once it is buried. This may have relevance once the body is unearthed and an attempt is made to reconstruct the sequence of past events.

Method

Bodies or meat samples are infested with blowfly maggots for varying lengths of time and then weighed and buried at different depths. After set periods of time, the samples are dug up, reweighed and the number of live maggots, their stage of development and the extent of decomposition are recorded and compared to control unburied samples.

Title

Is there a difference in the development rates of blowflies from different parts of the country?

Rationale

The accuracy of minimum time since death calculations based upon the age of blowfly larvae depends upon the reliability of laboratory data on larval development rates. However, there is little published evidence on whether there are major differences in the development rates of regional strains, especially those living under very different climatic conditions.

Method

Blowflies of the same species would be collected from the wild in the extreme north of the country and the southern regions and their development rates under identical conditions compared to those of a laboratory strain. The experiment would include determining the minimum temperature at which hatching and development would occur and the effect of exposure to short periods of high temperature, such as would occur in a maggot feeding ball.

Title

Does food processing affect the morphology of the insect surface cuticle?

Rationale

People often complain of finding insects in tinned and packaged goods but it is not always certain whether they were added after the packet was opened or they fell in at some stage in the processing.

Method

Cockroaches, tenebrionid beetles or adult blowflies would make suitable experimental animals. The insects may be killed using carbon dioxide and the morphology of the cuticle observed after varying treatments such as boiling, cooking in a pressure cooker, etc. A scanning electron microscope will reveal the best detail.

8 Vertebrates in forensic science

Chapter outline

Vertebrate scavenging of human corpses
Vertebrates causing death and injury
Neglect and abuse of vertebrates
Vertebrates and drugs
Vertebrates and food hygiene
Illegal trade in protected species of vertebrates

Objectives

Identify the vertebrates most commonly responsible for dismembering and consuming corpses and describe how to distinguish between damage caused by scavengers from pre- and postmortem injuries inflicted by humans.

Explain the circumstances under which vertebrates cause death and injury. Discuss the techniques that can be used to link a vertebrate with the scene of a crime.

Explain the difficulties of proving cases of neglect and abuse of domestic and wild animals and how childhood cruelty to animals may be linked to violent behaviour in later life.

Describe the illegal use of drugs in racing and how vertebrates are used to import contraband drugs.

Differentiate between valid and fraudulent claims of food contamination.

Discuss the extent of the illegal trade in protected species of wild animals and wild animal products and describe the techniques that can be used to identify the provenance of an animal or its body parts.

Essential Forensic Biology, Alan Gunn
© 2006 John Wiley & Sons, Ltd

Introduction

Vertebrates (fish, amphibians, reptiles, birds and mammals) are nowhere near as numerous or as diverse as invertebrates but their larger size (usually) and, in the case of mammals, shared characteristics with ourselves tend to make them the first thing we think of when asked to 'describe an animal'. In forensic biology, vertebrates tend to be as much the victims of crimes as sources of evidence that can be used to solve them and police forces often have special units assigned to deal with such incidents. For example, in the UK, the Metropolitan Police Force has a Wildlife Crime Unit based within New Scotland Yard (www.met.police.uk/wildlife/new%20site%20docs/docs/wcu.htm) that works in partnership with police forces at both a national and international level.

Vertebrate scavenging of human corpses

Many vertebrates will exploit human corpses as a source of food. In Northern European terrestrial ecosystems this usually means dogs and other caniids, rats, pigs and birds such as crows, ravens, buzzards and jackdaws, whilst in aquatic ecosystems various fish and seagulls are responsible. However, even vertebrates that are normally considered to be herbivores, such as squirrels (red and grey), sheep and cows, will gnaw on bones, especially if they are living in a nutritive-poor environment (Figure 8.1). Dogs are well known for their scavenging activities and will spread body parts over several metres. Birds will also cause the scattering of remains and may remove small body parts to their nests. Dogs frequently carry away and bury bones and limbs they find and this may result

Figure 8.1 Bones gnawed by a grey squirrel. Note the characteristic paired grooves caused by their incisors. Many rodents gnaw on bones either to sharpen their teeth or to reach the nutritious marrow

in it being impossible to recover a whole skeleton. Badgers are reportedly partial to feet and will take them away for later consumption in their set, whilst porcupines are notorious for collecting all sorts of objects in their burrows ranging from bones to tin cans. Experiments with hyenas and leopards (Pickering, 2001) indicate that fingers and toes may pass through their digestive system and be found in their faeces, and there is a lot of information in the archaeological literature on the analysis of bone fragments in the scats of foxes and other carnivores. Rings and other jewellery, especially gold, will pass through a digestive system largely unharmed whilst bones that have passed through the digestive system usually show signs of acid etching. Therefore, an analysis of nearby faecal deposits and any dens, sets or burrows may yield missing bones and jewellery. Skeletons of young children do not survive a scavenging attack as well as those of adults. Not only are their bones smaller and weaker, but also the epiphyseal plate, a band of proliferating and developing cartilaginous cells in the epiphysis (head) of the long bones, is thicker in children and is easily chewed away by a carnivore. The bone shafts can then be swallowed and will be broken down more readily in the stomach because of their lower calcium content. Similarly, the sutures between the skull bones of infants and young children are movable, thereby enabling the skull to be broken more easily than that of an adult. Because of their lower calcium levels, children's bones will also decompose more readily in acid soils.

The ability of dogs to locate human remains even after burial sometimes results in a body being unearthed despite the best efforts of a murderer to conceal it and this has been exploited in the training of so-called 'cadaver dogs' by police agencies (Lasseter et al., 2003). Scavenging invariably results in serious damage to the victim's body, up to and including decapitation, and may lead an investigator to assume initially that a person who died of natural causes or suicide was the victim of a brutal killing. In addition, dogs (especially) will tear off and scatter the clothes from a dead body and there are several reports of them removing the genitals/genital regions of both men and women, thereby raising the suspicion of sexual assault (e.g. Romain et al., 2002). The damage that scavengers inflict therefore needs to be distinguished from that caused at the time of death or by a murderer cutting up the body of their victim. Dogs and other vertebrate animals inflict characteristic tooth marks on bones that can be easily distinguished from the cutting or sawing damage caused by human tools and they do not induce bevelling or concentric or radiating fractures such as those caused by gunshot wounds or trauma induced from a blunt or sharp implement. Archaeologists have done a lot of work on how to distinguish between the damage caused to bones by different animals, as opposed to humans, and also the effect of different types of cooking (e.g. Brain, 1981). On soft tissues, tearing and puncture marks can sometimes be matched with the tooth structure, dental formula or claws of the animal responsible, and the presence of animal hairs in the wound and/or faeces in the vicinity provides further corroborating evidence. Dogs and other caniids tend to spend a long time chewing on a bone, all the while turning it over and over, resulting in a mass

Figure 8.2   Bone gnawed by a dog. Note the mass of grooves and pits that make it difficult to discern individual tooth marks. Scale in millimetres

of grooves and pits from which it is difficult to discern individual tooth marks (Figure 8.2). By contrast, domestic cats do not cause as much physical destruction and their bite marks tend to be more dispersed and the tooth marks more defined (Moran and O'Conner, 1992).

Bite marks inflicted after death – like all such wounds – do not bleed to any great extent and the characteristic tooth marks may enable not only the species responsible to be identified but also the individual animal. The latter is especially the case where the person has died indoors or in an enclosed space in which the suspect pet(s) or domestic animal(s) is also confined. In addition to the tooth marks, animals may also leave hairs from their muzzle whilst feeding, the individual characteristics of which, along with extracted DNA, may be used for identification. The stomach contents of pets and domestic animals may also be examined to determine whether they had been feeding on a body but it is seldom possible to catch wild or feral animals for analysis. In addition to physical evidence, it is important that witness statements should be obtained (where possible) of the dog's (or other animal's) past behaviour, its behaviour at the time it was impounded and an assessment made by an experienced animal handler or behaviouralist over the following days to determine its mental state and relationship to humans. In well fed and otherwise normally behaved pets, postmortem feeding damage caused by confined dogs and cats is commonly thought to take place long after death has occurred and be induced by the absence of any other suitable food source. However, there is probably a great deal of variation between cases and Rothschild and Schneider (1997) have described one in which scavenging by an Alsatian dog took place within 45 minutes of the owner committing suicide by shooting himself in the mouth (hence a large wound was already present on the body). They discussed several possible explanations for the early onset of scavenging, including the aggressive behaviour caused by being confined and the dog, being a pack animal, attacking its owner at a time of weakness in an attempt to gain social domination. However, they considered the most likely explanation in this case was that the dog attempted to help its unconscious or recently deceased owner, first by nuzzling and licking and then when these failed to become panicked into attacking and mutilating him.

Rats will inflict damage before death has occurred and historical accounts of soldiers and prisoners living in unhygienic circumstances often mention rats nibbling at fingers and toes. They also favour soft, moist areas such as the eyelids, nose and lips. Consequently, rodent damage may occur both before and after death in a person who has died of wounds, disease or intoxication over a period of days and is not unusual on the bodies of homeless people or those living in squalid conditions. Large numbers of rats are capable of overwhelming and killing a person who is already comatose or too weak to defend himself but documented cases of such instances are very rare. Unhygienic circumstances are not always a factor in rodent scavenging because many people keep rats and other rodents as pets. For example, Ropohl *et al.* (1995) have described postmortem damage caused by a free-range golden hamster that was so extensive that it was initially believed to be the work of a murderer attempting to scalp his victim. The hamster was easily identified as the culprit, because of its typical 'rodent signature' – rodents often leave characteristic faecal pellets (their shape and size varies between species) whilst feeding and their paired chisel-shaped incisors cause crater-like lesions with notched edges in soft tissue (Tsokos and Schulz, 1999; Tsokos *et al.*, 1999). The hamster further incriminated itself by taking fragments of skin and tissue back to its nest – another typical rodent trait. Bite marks do not always cause tissue loss and by stretching the skin it is sometimes possible to see marks caused by the paired incisors. Where rodents have gnawed on bones they leave paired parallel grooves with intermediate 'groins' – the width of the grooves indicates the size of the incisors and hence the size and probable identity of the rodent species.

In mediaeval times, pigs used to be allowed to roam freely and there are reports of them biting and even killing and eating babies. Nowadays this is no longer a problem in Northern Europe although the wild pigs and boar that can be found in some regions would probably be happy to exploit any dead bodies left in their woods. During experiments with domestic pigs that were fed fresh uncooked bones of sheep, cattle and pigs, Greenfield (1988) found that sows tended to briefly chew on the first bone they encountered before dropping it and moving on to another and repeating the process. After a short period of time the pigs concentrated on the smaller bones, especially the vertebrae, which could be picked up and carried around, and these were completely consumed. Large bones such as the femur were damaged, especially at the ends, but were not totally destroyed. Crime writers sometimes suggest that a good way to dispose of a dead body would be to feed it to pigs but this requires the cooperation of the pig farmer because a whole body will not be entirely consumed overnight and it is doubtful that the pigs would be able to deal with all the bones.

Birds such as crows frequently begin feeding on the eyes and, if extended, the tongue and over time they can cause sufficient tissue loss to make the cause of death difficult to determine (Asamura *et al.*, 2004). As sheep farmers will testify, birds do not always wait for an animal to die and any creature too weak to defend itself may be attacked. Similarly, sailors who have had to abandon ship and end up swimming in the sea have sometimes stated that gulls and albatrosses have attacked them and that some people have drowned in their attempts to avoid the birds. The lack of eyes and the presence of head wounds may therefore be a consequence of birds rather than human activity and may be caused either before or after death has taken place. Birds tend to produce stab-like wounds, the size and depth of which vary with the size of the bird's beak. Unless birds have been seen feeding on a body and/or they have left their faeces, it may be difficult to implicate them with the damage caused – although the lack of bleeding would indicate that it took place after death. However, bleeding may be extensive in the case of a body floating in water (see Chapter 1).

Many species of fish, both freshwater and marine, will feed upon dead animals but there is little information available on their importance in forensic investigations.

Vertebrates causing death and injury

Domestic and wild animals are seldom a cause of death in northern Europe but such cases generate enormous publicity and fear. By contrast, dog bites are common and may result in serious injury – and consequent litigation. There are intriguing differences in the number of reported dog bites between countries and these probably reflect national attitudes towards the keeping and training of dogs. For example, Vasiliev *et al.* (2003) cite a figure of 200 000 attacks per annum in the UK but only 30 000 in Germany; figures as high as 5 million attacks per year have been cited in the USA (www.dogbitelaw.com). Wounds are usually caused to the extremities, especially when the attack comes from stray or feral dogs, whilst pet dogs are reportedly more likely to attack the face or neck. The wounds may be life-threatening if major blood vessels such as the femoral artery are damaged. The wounds may also become infected with bacteria transmitted in the dog's saliva and in some countries there is always the risk of rabies. In fatal cases, there is serious loss of blood and often damage to the hands and arms where the victim has attempted to ward off the dog(s). Young children and the elderly and infirm tend to be those most at risk. Domestic cats have finer and sharper teeth than dogs that enable them to penetrate bone despite their comparatively weaker bite. Puncture wounds occur in 57–86 per cent of cases of cat bites, lacerations in 5–17 per cent and superficial abrasions in 9–25 per cent of cases (Dire, 1992). Cat owners commonly have fine

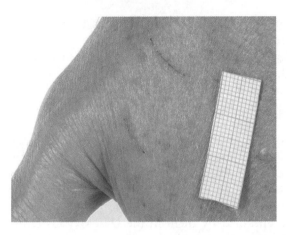

Figure 8.3 Cat scratch marks, in this case partially healed, should not be mistaken for evidence of assault although they do indicate that the person may have owned a cat. Scale in millimetres

scratch marks (Figure 8.3) on the backs of their hands as a consequence of playing with their cats. Cats seldom attack unless they are cornered and feel threatened, but when they do they can cause serious lacerations before escaping as quickly as possible. There are no reports of them causing death by attacking someone although tripping over a circling cat is a common cause of injury, which may be serious. For example, an elderly person may be found unconscious with a head injury and crime may be suspected initially until other evidence – such as the pattern of blood spatter – indicates a fall (although if the person does not recover or cannot remember what happened, this would be difficult to distinguish from a push by an intruder).

Dogs are sometimes used to intimidate or injure victims and in Germany, where it is illegal to carry firearms without a license, large aggressive dogs are reportedly popular with skinhead gangs and drug dealers. Similarly, the use of guard dogs to intimidate prisoners has become notorious in certain American prison camps in Iraq and Afghanistan. Proof of intimidation is very difficult to establish if no physical injuries have been caused, although the American prison guards were callous (and stupid) enough to photograph themselves committing the crimes. In addition to using dogs to bite and intimidate victims, some people have trained them to commit sexual acts and there are reports of them being incited to commit rape (Vintiner *et al.*, 1992; Schudel, 2001). Consequently, victims of sexual assault may have traces of non-human spermatozoa and it may require modifications to existing protocols for their detection.

Dogs are not the only animals that may be used to intimidate people and in 2002 there was an unusual case of a woman throwing her pet iguana first at a pub doorman and then at a policeman. Fortunately, neither lizard nor humans were harmed in the incident but the owner was charged with causing the animal unnecessary suffering. Iguanas can grow to several feet in length and, if provoked, are capable of biting off a man's fingers. Even dead animals can pose a risk when in the wrong hands. For example, in August 2005, a man found himself in Dudley Magistrate's Court as a consequence of hitting his wife with a 3-foot long pike. According to his wife 'John has got into a temper before but he has never attacked me with a fish'. The man escaped with a six-month conditional discharge, his wife forgave him and the fish was cooked and fed to their pets (The Daily Telegraph, 5 August 2005)

In addition to the risks posed by normal pets and domestic animals, the opportunity and increasing popularity for keeping exotic animals has resulted in more people coming into contact with large and potentially dangerous creatures (Lazarus et al., 2001). For example, there are currently more tigers in captivity than there are in the wild and during 1998–2001 27 persons were injured and a further seven killed by them in the USA (Nyhus et al., 2003). Some of the deaths caused by captive wild animals are the result of people taking inadequate precautions when housing or handling them – these are usually animals that are kept by private individuals. Consequently, in the UK, alterations are being proposed to the Dangerous Wildlife Act 1976 (www.defra.gov.uk/wildlife-countryside/gwd/wildact.htm) under which animals deemed to be dangerous must be licensed and kept under strictly regulated conditions – this does not apply to zoos, pet shops, circuses or registered scientific establishments because these are covered by their own specific legislation. To keep any animal listed under the act, a private individual must obtain a license from their local authority. The owner must have insurance, the animal must be kept under appropriate conditions and its movements from the premises are governed by a series of restrictions. Local authorities will have the right to enter premises if they believe that animals covered by the act are being kept there illegally. The keeping of exotic pets is usually a difficult, expensive and time-consuming occupation and when the animals grow too large or aggressive it is not unusual for them to be dumped. For example, iguanas have become popular pets in the UK, where they are usually purchased when they are a few inches long. However, when mature they may measure up to 5 feet and can become aggressive – at which point they are looked on as a burden. The killing or dumping of unwanted exotic pets can lead to criminal charges although proof is difficult, especially if the animal was being kept illegally. Zoos have a duty not only to maintain potentially dangerous animals under safe conditions from which they cannot escape but also to enable them to be seen by members of the public whilst simultaneously preventing the naïve, deranged or suicidal from coming into contact with them. For example, climbing into a lion enclosure is sometimes used as a form of suicide (e.g. Bock et al., 2000).

Identifying a dog or other animal responsible for an attack or for causing an accident is not always easy, especially when it runs away afterwards. Most people, when asked to describe a dog, would be unable to provide more than vague identifying features and statements such as 'it was a large brown Labrador and the accused is also a large brown Labrador' are not going to be much use as evidence in a court of law. Animal hairs are characteristic of the species and can also provide information on coat coloration but their presence provides only weak evidence of an association with any individual animal. Dogs and cats are two of the most popular pets in many parts of the world and, as any owner will agree, pet hairs tend to get everywhere despite strenuous efforts to keep rooms and clothing clean, so they are a potential source of forensic evidence (D'Andrea et al., 1998). Animal hairs vary in length and colour between different parts of the body and this should be borne in mind when collecting specimens. Animal hairs may also originate from a fur garment or pelt – these hairs are often coloured, trimmed and lack a root. Just as pets and other domestic animals transfer their hairs onto humans, so humans transfer clothing fibres onto their pets and brushing their fur or coat can yield evidence of contact. However, even when matches are found, the evidence remains circumstantial. Consequently, increasing use is being made of DNA technology (e.g. Savolainen and Lundberg, 1999; Pfeiffer et al., 2004). Schneider et al. (1999) describe a case in which a dog that was believed to have been responsible for causing a traffic accident was exonerated by comparing the sequence analysis of the mitochondrial D-loop control region of hair fragments found on the damaged car with those of samples obtained from the suspect dog. However, they found that this technique was only suitable for excluding suspects rather than identifying culprits owing to limited polymorphism of the canine mitochondrial D-loop region. Similarly, Brauner et al. (2001) were able to exclude a dog suspected of mauling a young girl by comparing micro-clots of blood found on the dog's coat with the girl's blood using STR (short tandem repeat) DNA analysis. As usual, this evidence was not used on its own but was additional to hair and fibre analysis. It was impossible to perform a meaningful comparison of the dog's dentition with the bite marks – this is a useful reminder that evidence should be collated from as wide a variety of tests and sources as possible because, depending on the circumstances, they vary in their applicability and sensitivity. According to Eichmann et al. (2004), canine STR typing is more effective when swabs are taken from severe bite wounds than those that are relatively light. At first sight this appears odd because severe wounds bleed heavily and therefore the swabs would be badly contaminated with human DNA. However, severe wounds result from extremely forceful and often prolonged contact between the dog and its victim: consequently the dog's saliva is transmitted liberally into the wound and smeared onto the surrounding skin. By contrast, light wounds usually result from a quick snap or nip resulting in relatively little saliva being transmitted and, crucially, these wounds are more likely to be washed before medical attention is sought, thereby further reducing the amount of canine DNA present. Analysis of DNA from dogs and other domestic animals is not only of value when the animal itself is the suspect or victim of a crime but may also be used as a means of providing a link between people or between people and a location. For example, dogs and cats are two of the most common pet animals and many of us unintentionally (and often unwillingly) carry with us evidence of that association. The techniques for the extraction and forensic analysis of DNA from domestic animals lag behind those for humans and the hairs themselves can present particular problems (Fridez *et al.*, 1999). For example, owing to the small quantities of DNA present in a single dog hair, it can be difficult to DNA-type samples unless there are ten or more 'good quality' hairs (i.e. complete and with the roots still present) available (Pfeiffer *et al.*, 2004). Individual animals have their own unique smell that is easily transferred to clothing – as evidenced by the keen interest shown by one's pet cat or dog on returning home having stopped to stroke another animal along the way – but there is currently no means of identifying animals in this way for forensic purposes. Humans also have their own individual unique smell and although there are now electronic sensors, such as the zNose (http://www.estcal.com), that can characterize odours, dogs remain the most sensitive smell detectors and can be trained to detect chemicals ranging from cannabis to petrol. The possibility of recovering oral bacteria from animal bite marks and using these to identify the animal responsible for the bite does not appear to have been investigated, although it is showing promise in the examination of bites inflicted by humans (see Chapter 2).

Neglect and abuse of vertebrates

Whilst pets and domestic animals sometimes attack and may even kill humans, they are themselves far more frequently the subject of neglect and wilful abuse that may result in the animal dying. Wild animals are also frequently killed for no other reason than personal gratification. To mistreat animals in this way has long been recognized as a crime but identifying such activity has received a higher profile following the realization that childhood cruelty to animals is often linked to the development of violent behaviour towards humans in later life (Raup, 1999; Lockwood, 2000; Dadds *et al.*, 2002). The diagnosis of neglect and abuse is a job for veterinary surgeons based on the clinical symptoms, and where the animal is voluntarily brought to the surgery they face the dilemma of reporting the owners to the police and thereby risking the animal suddenly 'disappearing' before it can be impounded or keeping quiet and attempting to treat the animal whilst encouraging the owners to behave more responsibly. Because the mistreatment and illegal killing of both wild and domestic animals can result in hefty fines and imprisonment, the person charged can be expected to mount an active defence and is likely to escape punishment on a legal technicality if the investigation has not been conducted according to a recognized procedure and to the same standard as the forensic examination of a human crime victim (Cooper, 1998). In order to improve the success of prosecutions for badger baiting, the RSPCA is currently funding a scheme to establish a national database of badger DNA. This will enable investigators to confidently identify traces of blood, fur and faeces found on a defendant's clothing, vehicle or dogs. Sadly, as in human cases, good practice and reality are not always the same. This was supremely exemplified in the McMartin Preschool child molestation and animal sacrifice case – 'the longest and most expensive trial in the history of Los Angeles, California' (Brattstrom, 1998) – in which all the charges were proved to be false. Amongst the many failings in the collection and interpretation of 'evidence' was a box containing material that was obviously 25–50 years too old, the presentation of undamaged turtle bones of an alleged animal victim when it is extremely difficult to kill a turtle without considerable violence, and damage to long dead specimens that had almost certainly been done during clumsy attempts at unearthing them (Brattstrom, 1998).

Humans have always used a wide range of both vertebrate and invertebrate animals for their own sexual pleasure and Ancient Greek literature is full of strange couplings between humans and other animals – often claimed to be gods in disguise, which is a pretty feeble excuse. However, such activities were frowned on by most communities and if discovered could result in judicial proceedings in which both the man (it usually was a man) and the unwilling object of his attentions were condemned to death. Bestialism was a capital offence in the British Isles as late as the nineteenth century: John Leedham had the unfortunate distinction of being the last person in Derbyshire to be hanged for a crime other than murder when he was executed for bestialism outside the Derby New County Gaol on 12 April 1833. It remains an all too common and under-reported problem, probably as a consequence of its somehow simultaneously ludicrous and unpleasant nature. In a survey of small animal veterinary practitioners in the UK, Munro and Thrushfield (2001) found that 6 per cent of 448 reported cases of non-accidental injury were of a sexual nature. As mentioned above, such activities are not only of concern for the distress caused to the animals but also for the possibility that they might lead to assaults on humans – although research is still needed in this area.

Wild animals do not always kill their prey quickly and cleanly, and neither, contrary to popular belief, do they kill only sufficient to assuage their appetite. Consequently, one may find badly wounded wild or domestic animals that at first sight appear to have suffered at the hands of a sadistic individual. For example, in April 2004 numerous dead and dying frogs and toads were found in Aberdeenshire with their hind legs ripped off and this sparked a police investigation. It was subsequently discovered that the culprits were otters, which bit off the hind legs of their victims and then skinned them. In the case of toads, this was done to remove their poison glands.

Vertebrates and drugs

Dogs have a keen sense of smell and this has led them to be used by police forces throughout the world to detect hidden drugs and explosives at airports, train stations and ferry terminals. Intriguingly, African Giant Pouched rats (*Cricetomys gambiensis*) have a similar ability to be trained to identify distinctive smells in return for a reward. At present, they are being assessed for their ability to detect tuberculosis in sputum and explosives in landmines (www.gichd.ch/pdf/publications/MDD/MDD_ch4_part2.pdf) but there seems no obvious reason why they could not also be trained to detect contraband drugs. The rats are said to be 'more mechanical than a dog and they are easier to transfer to different owners'.

In addition to detecting drugs, animals may also be used to smuggle them – although the extent to which this is happening is not known. For example, in September 2003 at Schiphol airport, The Netherlands, two Labrador dogs in transit from Colombia were discovered to have a total of 21 packets of cocaine sown into their stomachs. Suspicions were raised by the dogs' behaviours, one being aggressive whilst the other was weak. Both dogs had scars and X rays revealed the presence of the containers. One dog survived the removal of the containers whilst the other had to be put down because they had fused to the stomach lining. Two persons were arrested when they arrived to pick up the dogs but the Colombian authorities have yet to apprehend the exporters and the veterinary surgeon who implanted the drugs. The presence of scarring and the unusual behaviour of any animal being moved between countries should therefore arouse suspicions.

Just as the use of drugs is a problem in human sporting activities, so is the administering of drugs to competition animals. The true extent of the problem is uncertain. The University of Ghent, Belgium, runs a doping control laboratory and, according to their data between 1993 and 2003, the percentage of horse samples testing positive varied considerably (1.2–8.4 per cent) whilst the percentage of human samples testing positive remained fairly constant (3.6–6.6 per cent). Testing is now routine at many events and in the UK there is routine random testing of racehorses. Testers are also authorized to arrive at stables unannounced to collect their samples. The drugs involved are often those used in the treatment of disease but are being administered solely to improve performance: examples include erythropoietin (EPO, which increases the red blood cell count), clenbuterol (a bronchodilator) and propantheline bromide (blue magic, a muscle relaxant that also acts to increase a horse's lung capacity). Consequently, owners of competition horses have to be extremely careful over the medication their animals receive and to be aware of the risks of their animals receiving spiked feed or being inadvertently fed inappropriate food. For example, in America, owners of show horses are warned not to allow their children to reward them with drinks of coke because it might increase the caffeine levels above allowable levels. The use of 'downers' to reduce the activity of horse is also alleged to be common practice, although supporting evidence is not readily available. This might be done to make a troublesome horse more placid at the time of sale, to cause it to lose a race or to make it more manageable in the ring. Ponies UK has begun a programme of random dope testing following judges and stewards voicing concerns over the placid behaviour of some of the smaller horses ridden by children. Bute (phenylbutazone), a non-steroidal anti-inflammatory drug used in the treatment of strains, sprains and feverish symptoms, is one of the suspects although non-specified herbal remedies are thought to be involved. In 2002, a British trainer was fined £600 by the Jockey's club when one of his horses was found to contain traces of the 'stopping' drug acetylpromazine (ACP). Obviously the drug was not in a high enough concentration because the horse won by 11 lengths.

Vertebrates and food hygiene

Most food hygiene litigation involving vertebrates relates to rats and/or mice gaining access to stored food and damaging it through physically eating it and contaminating it with their faeces and urine. Even if the animals are not seen, their faeces have a characteristic shape and size that enables identification. For example, the faeces of brown rats (*Rattus norvegicus*) are usually deposited in groups (although it may be scattered) and tend to be spindle shaped, whilst the faeces of black rats (*Rattus rattus*) are scattered around and tend to be sausage or banana shaped. Food hygiene litigation tends to involve food hygiene inspectors rather than forensic scientists: there is seldom an issue over when the contamination occurred or who was responsible. For example, if rat faeces were found in the kitchen of a restaurant, rats must have gained access and it does not matter when they did or how many of them did, or where they came from, it is the proprietor's responsibility to ensure that they do not enter the building and he/she will be prosecuted.

Anecdotal reports of finding the tails of mice or rats in food products are commonly heard but many of these are probably apocryphal. They do, however, sometimes encourage people to attempt to falsely sue a food retailer by claiming to have found a rat or mouse, or a bit of one, in their food (e.g. Platek *et al.*, 1997). Williams (1996) describes such a case in which a man stated that having consumed a can of commercially produced milk stout he found a dead mouse at the bottom of the can. Williams first determined that it was possible for the mouse to have gained access via the can's ring pull and then carried out a series of experiments in which dead mice were placed in cans of milk stout, which were then sealed on the factory production line. Some of the cans were then pasteurized (as normal) and others were not. Both sets of cans were then stored under ambient conditions or refrigeration and opened after set periods of time. In all cases, the experimental mice were found to have undergone considerable decomposition within 1 month and after 3 months they were completely disintegrated. By contrast, the mouse obtained from the complainant's can, which was 3 months old, was well preserved and exhibited minimal postmortem changes. It was concluded that the complaint was fictitious. This case emphasizes the importance of carrying out carefully designed experiments to exclude all possibilities – for example, the effect of pasteurization.

'Bushmeat' is the term given to the meat of wild animals that has been caught by villagers in central and Western Africa and sold in the local market. For many years, the practice was 'low key' and was not thought to cause any harm to the local wildlife population – indeed, it was thought by some to be beneficial because it made the local people more aware of the need for conservation. However, in recent years there has been a large increase in the trade, partly owing to increased demand and partly through greater access to the forests via logging practices. It has become highly profitable: a single suitcase of bushmeat may have a street value of up to £1000 in the UK. Consequently, illegally imported bushmeat is being seized with increasing frequency at airports on its way to African communities in the UK and elsewhere. In 2003, an average of 427 kg per week of animal products, including bushmeat, were being confiscated at London's Heathrow airport. Some of the bushmeat being imported includes that of protected species and, much of it is not being transported correctly, thereby posing a risk of food poisoning, and there is now a realization that it is a potential source of anthrax (Leendertz et al., 2004) and zoonotic viral infections (Peeters, et al., 2002).

Illegal trade in protected species of vertebrates

As the wild populations of animals decline in many parts of the world, usually owing to mankind's activities, the need for effective conservation measures grows. To be truly effective, these measures require to be backed up with laws to ensure their enforcement and, consequently, litigation follows. On an international scale, wild animals are offered varying levels of protection under the Convention on International Trade in Endangered Species (CITES: www.cites.org/). Those species listed under Appendix I are fully protected and no trade is allowed, whilst those listed under Appendix II can be traded under a permit system. Individual countries also have their own laws that control which animals can be traded and hunted – for example, the shooting of songbirds is legal in some parts of Europe but illegal in the UK.

Three of the most common problems are the unlicensed trade in live animals, the unlicensed trade in body parts (skins, bones, meat, etc.) and the intentional killing of animals for sport or because they were perceived as a threat. There is a global market in live wild animals and their body parts that, for the most part, is perfectly legitimate and can provide employment to poor people in developing countries. However, as the animals become scarce, their value increases and people are tempted into criminal activity – the illegal sale of plants and animals is estimated to be worth billions of pounds per year on a worldwide basis. In response, the laws have become tougher and in the UK the maximum prison sentence for illegal trading in wildlife has increased from 2 to 5 years. Forensic science becomes involved at a variety of levels, e.g. in determining the provenance of an animal, identifying bones or animal parts and determining the cause of death.

Identifying the provenance of an animal is important because in the UK, provided that one is a registered breeder/seller with a license from the Department of Environment, Food and Rural Affairs, it is not illegal to sell an endangered species if it has been bred from existing UK stock. For example, tortoises and parrots were once common pets but since it became illegal to import them from wild populations abroad their numbers here have declined and their value has increased. Proof of provenance usually depends on the presence or absence of the appropriate documentation, although the use of DNA technology can prove useful in disputed cases. For example, there is a highly profitable and illegal trade in falcons and other birds of prey for use in falconry both within the UK and to supply the Middle Eastern market. Birds that are born and bred in captivity may be traded legally but these are insufficient to meet the demands. In one case the Metropolitan Police Wildlife Crime Unit was able to successfully prosecute a gamekeeper by using DNA technology to prove that the birds were taken illegally from the wild.

All species of wild birds, as well as their nests and eggs, are protected under British Law, although allowances are made to control pest species, such as feral pigeons, and for the shooting of game birds. Egg collecting was once a common hobby, especially amongst schoolboys. However, in the UK it became illegal to collect the eggs of wild birds in 1954 and since 1982 it has been an offence to even own the eggs of wild birds. Despite this, egg collecting remains a hobby of some people and they can have a serious impact on the populations of the rare species, such as many of the birds of prey. The eggs of such birds sell for high prices on the 'underground market'. Proving the provenance of wild bird eggs is less of a problem in ensuring a conviction because possession alone is a crime – although it always helps to know when and where the eggs were obtained. Wild finches and other songbirds are sometimes illegally trapped for sale as caged pet birds both within the UK and in the Mediterranean regions or to supply the gourmet food trade. As mentioned above, proving the provenance of these birds depends on documentation and, sometimes, DNA analysis. Although there is a legal trade in captive bred finches, those that are caught in the wild are thought to posses brighter plumage and therefore command a higher price and will sell for up to £100 each.

Farmers and gamekeepers sometimes attempt to kill eagles and other predatory birds and mammals – this is usually done by shooting or providing poisoned baits – because they believe that they kill lambs and game birds (Stroud, 1998). Therefore, the finding of a dead eagle, for example, especially one in outwardly good condition, might be considered suspicious. It is the responsibility of a vet to carry out a careful autopsy and submit tissues for toxicological analysis. If the evidence is to be used in a court of law, the autopsy must be performed with the same attention to detail as that of a human, with careful record keeping to ensure a 'chain of evidence' (Munro, 1998). Sometimes the wrong animal becomes an unintentional victim of a poisoning campaign. For example, on Crete, there has been a decline in the number of Eleonara's falcons that has been linked to the carbamate insecticide methomyl, which is sold under the trade name Lannate. The birds were not being poisoned by spray drift, or from consuming treated insects – the birds feed on the wing and poisoned insects are rapidly paralysed and therefore cannot fly. Instead, the birds were being poisoned as a consequence of drinking from clay bowls of water that had, illegally, been laced with the insecticide. Local farmers leave such bowls out to kill wild animals that are believed to be attacking their crops. As a result, the manufacturers of the insecticide are planning to add an unpleasant-tasting substance to the formulation to prevent this practice. A full description of the detective work involved in this project is available at www.birdlife.net (release date 31 May 2001). Sometimes it is important to determine the minimum time since death of a wild animal and this can be done using the same entomological evidence techniques as those detailed earlier. Anderson (1999) has described a case in Canada in which insect evidence from two illegally killed black bears was used to establish the fact that the bears had died during the time the suspects were in the locality and this information helped in securing a conviction.

Animal body parts from protected species are sometimes intentionally mislabelled or wrongly described in an attempt to avoid detection whilst at other times they are stated as being present when they are not. Both actions are undoubtedly illegal although the former has the most deleterious impact on the population of the protected species. A classic case of this is the use of tiger bones in traditional Chinese medicine. Literally hundreds of patent remedies claim to contain tiger bones even though China has agreed to ban their use. Wetton et al. (2002) have developed a highly sensitive tiger-specific real-time PCR assay capable of detecting fewer than ten substrate molecules from tiger bone, blood and hair but their analysis of a range of 'tiger bone' samples from traditional Chinese medicine shops found that in all cases the bone fragments actually belonged to pigs or cows. Their inability to detect tiger components indicated widespread substitution or their use at much lower levels than were being claimed on the packaging. Wetton et al.'s methodology should prove extremely useful in detecting the presence of body parts from other protected species in food and medicines. Rhino horn is another commonly used ingredient in Chinese traditional medicine and whilst the horns themselves can be easily identified from their morphology it is more difficult once they have been ground to a powder or made into sculptures. By amplifying and then DNA sequencing a partial (402 base pair) fragment of the cytochrome b gene it has been shown possible to distinguish between species of rhinoceros and to detect the presence of rhinoceros DNA even when powdered rhino horn is diluted with cattle horn (Hsieh et al., 2003). Sequencing the whole cytochrome b gene (1140 base pairs) is not usually possible in processed rhino horn because the DNA is degraded. In a similar way, the use of molecular phylogenetic analysis using mitochondrial DNA has proved useful in determining the origin of meat sold in shops – by which time processing has removed many of the normal identifying features and the nuclear DNA has become degraded to a greater or lesser extent. Using this approach, Marko et al. (2004) found that 77 per cent of the fish sold as red snapper were actually other species – such widespread misrepresentation has serious consequences for the management and conservation of fish stocks.

Molecular techniques are not the only ones that may be employed in determining the origin of body parts. For example, X-ray diffraction can be used to distinguish rhino horn from that of other species and ivory from antlers (Singh and Goyal, 2003) whilst isotope ratio analysis can be used to determine the geographical origin of ivory and bone (Stelling and van der Peijl, 2003). The latter technique relies on the fact that the chemical characteristics of the soil on which an animal or plant lives are reflected in the levels of chemicals and their isotopes present in their bones and teeth. However, to be fully effective, this requires a comprehensive database of soil chemistry characteristics.

For a summary of the vertebrate animals and their forensic evidence, see Table 8.1.

Table 8.1 Summary of vertebrate animals and their forensic relevance

Animal	Forensic relevance
Dogs	Cause of human injury or death Cause of traffic accident Victim of neglect or abuse Use in illegal baiting or dog fighting 'Kidnapping' – pets, especially dogs, are sometimes stolen and held for ransom Doping (e.g. greyhound racing) Cause of postmortem damage to flesh and bones Source of DNA linking a person to a locality
Cats	Cause of traffic accident Victim of neglect or abuse Use in illegal baiting Source of DNA linking a person to a locality
Rats and other rodents	Cause of human injury Cause of postmortem damage to flesh and bones Food spoilage
Domestic livestock	Victim of neglect or abuse Rustling Fraud (e.g. illegal movements across borders)
Birds	Cause of postmortem damage to flesh and bones Victim of neglect or abuse Illegal trade in protected species Illegal killing of protected species
Fish	Cause of postmortem damage to flesh Fraud (e.g. mislabelling) Illegal trade in protected species
Wild mammals, reptiles and amphibians	Illegal trade in protected species Illegal killing of protected species Poaching

Quick quiz

1. State four ways in which the activities of scavengers may affect the evidence in cases of murder/suspicious death.

2. How would you distinguish between damage caused to bone by sawing and the gnawing of rats and dogs?

3. Briefly discuss three reasons why a pet dog might eat its dead owner.

4. How might pet hairs be useful in a forensic investigation?

5. Why is prevention of childhood cruelty to animals of relevance to criminal behaviour?

6. Why is random dope testing of horses becoming increasingly common practice?

7. What is 'bushmeat' and what are the concerns about its importation into the UK?

8. Briefly explain how the presence of tiger bones in traditional Chinese medicines can be demonstrated.

Project work

Title

Identification of incorrect labelling of meat products.

Rationale

Food products are sometimes fraudulently mislabelled. This can have especial significance for Muslims and Jews if pork meat has been added and for Hindus if beef is the
contaminant.

Method

Meat products such as sausages and meat pastes may be purchased and tested directly
or artificially contaminated to assess the sensitivity of the assay method. Pet foods
would probably yield interesting results because they are often a mixture of ingredients – for example, 'chicken chunks' cat food from various sources is often labelled
as containing only a 'minimum of 4% chicken'. The identity of the meat present in the
sample may be assessed using standard mitochondrial DNA analysis.

Title

The utility of cat and dog hairs as forensic evidence.

Rationale

Cat and dog hairs are commonly found on clothing and might be used to link a person
with a locality.

Method

Cat and dog hairs are obtained by combing from a wide variety of cats and dogs and
a record made of the part of the body they are obtained from. The hairs are then
analysed using a stage microscope to determine whether it is possible to identify individual animals from their hair characteristics. Sub-samples of the hairs would then be
subjected to DNA extraction and analysis to determine whether this improved the identification. Another approach would be to ask volunteers to visit the homes of cat and
dog owners, sit down, have a drink and then leave. Their clothes would then be
analysed to determine the extent and location of any pet hairs they had acquired. Preferably they would never have visited the house before, some householders would be
cleaner than others and they would or would not have come into contact with the pet.
One question would be whether it is impossible to visit a pet owner's home without
acquiring pet hairs on the clothing and whether certain clothing is more effective at
trapping pet hairs.

9 Collection of animal and plant material for forensic studies

Chapter outline

The importance of scientific rigour and safety issues when collecting biological material

Collecting and preserving diatoms and algae for forensic analysis

Collecting and preserving testate amoebae for forensic analysis

Collecting and preserving plant material for forensic analysis

Collecting invertebrates for forensic analysis

Killing and preserving techniques for invertebrates: hard-bodied invertebrates; soft-bodied invertebrates; invertebrate eggs

Invertebrate identification techniques

Collecting environmental information for forensic analysis

Objectives

Discuss common errors and risks associated with the collection of biological forensic data.

Explain how algae, and in particular diatoms, can be extracted from water and from tissues for forensic analysis.

Describe how plant material such as leaves and seeds should be stored.

Discuss how pollen samples should be sampled and processed to avoid the risk of contamination.

Compare the collection and preservation techniques suitable for different types of invertebrates.

Compare the different types of killing methods for invertebrates and explain why choosing the most appropriate method is so important for forensic specimens.

Discuss the benefits and drawbacks of traditional and molecular-based taxonomic methods of identification for forensic specimens.

Describe the various types of environmental data one should collect at a crime scene and explain why such data are important when analysing biological forensic data.

Essential Forensic Biology, Alan Gunn
© 2006 John Wiley & Sons, Ltd

Introduction

The information contained in this chapter relates primarily to homicide investigations but the basic procedures would be the same whenever biological material was required for forensic analysis. Basically, the collection of all evidence must follow a set protocol that ensures health and safety and the continuity of evidence.

The importance of scientific rigour and safety issues when collecting biological material

When collecting biological evidence there are recognized examples of good practice that must be followed, otherwise incorrect conclusions may be drawn or it may not be admissible in court. For example, the oldest maggot larvae may have been missed, pollen contaminants may have been introduced and the possibility of the mixing up of samples may not have been guarded against. Ideally, all samples should be collected from the crime scene by a trained person and according to a set protocol. If a body is involved, specimen collection should take place before this is moved and notes should be made at the time of any animals and plants present and their numbers, as detailed in the following sections. When a decision has been made to move the body, the hands should first be enclosed in clean plastic bags to preserve biological and chemical clues that may be attached to the palms or underneath the fingernails. If the feet are exposed, these should also, ideally, be enclosed in bags for the same reason. The bags also protect the skin from physical damage when the body is moved: during the later stages of decay, the surface of the skin can be sloughed off easily, and if the epidermal surface is damaged in the process this can make it impossible to take fingerprints. Afterwards, the whole body (still in its clothes, if present) should be placed in a clean body bag and removed to a mortuary where a thorough examination can take place. This ensures that evidence is kept together and reduces the possibility of contamination.

Health and safety should be a priority both at the crime scene and within the laboratory when examining the evidence. Extreme care should be taken when handling samples collected from corpses. Decaying bodies contain many bacteria and fungi, some of which may be pathogenic, and there is always the possibility that the victim may have been suffering from a transmissible disease. The risks extend beyond the tissues of the body because, for example, both anthrax and tuberculosis spores pass unharmed through the guts of blowfly maggots and adult flies so they too can be potential sources of infection. Wherever intravenous drug abuse is suspected to have taken place at a crime scene, special care needs to be taken when handling and storing exhibits such as needles owing to the risk of HIV and hepatitis B and hepatitis C. Illegal cannabis grow-rooms are seldom set up in accordance to building regulations and can therefore contain dangerous electrical wiring or present a fire risk. The analysis of pollen and diatoms usually requires the use of strong acids and should only be undertaken in a suitable laboratory following a COSHH assessment of the risks.

All biological specimens that are to be used as evidence should be accompanied by a card detailing the following information: its identification, the date it was collected, the precise geographical location and its position on the body or within a room (if appropriate), the name of the person who identified the specimen and a code to identify the crime with which the specimen was associated, along with an exhibit number. The information should be written in indelible pencil or permanent ink and attached so that it cannot become lost. For example, the card should be attached to the pin passing through the body of a mounted insect or placed within the tube containing specimens preserved in alcohol. Slide-mounted specimens should have a label with the same information affixed to them.

Collecting and preserving diatoms and algae for forensic analysis

Many species of algae, including diatoms, exhibit seasonal growth characteristics and some are notorious for forming short-lived toxic blooms, therefore if an attempt is to be made to match the species composition recovered from a body or object with a specific locality it may be necessary to take sequential samples through the year. In the case of open water, samples should be taken from both the surface waters and the bed; in the case of a river or stream, samples should also be taken both above and below the site at which the body was found because there may be differences in the diatom flora and the body may have been moved by the water currents. Owing to the small size of diatoms and their widespread occurrence, every effort should be made to avoid the possibility of sample contamination occurring in both the field and the laboratory. All collecting equipment (e.g. plankton nets, soil corers and collecting vials) must be scrupulously cleaned before use and within the laboratory the samples should be processed in clean glassware and laboratory procedures adopted to reduce the chances of contamination occurring (e.g. working within a laminar flow cabinet). Blanks, consisting of samples of distilled water, should be processed at the same time as a check to test whether contamination could be occurring.

Diatoms and other algae can be collected from open water using a plankton net – this consists of a long funnel-shaped net bag mounted on a circular frame and with a collecting vial attached to the narrower trailing end. The mesh size of the net will determine the size of the plankton catch – planktonic algae tend to be small and require a fine net size (0.1–0.3 mm). Algae and other microscopic organisms that are attached to underwater substrates are sometimes referred to as 'periphyton' and require specialized collection techniques. Haefner *et al.* (2004) have described a simple apparatus that consists of two syringes, A and B, coupled together. Syringe A has its tip sawn off so that the barrel can be directly applied to the substrate whilst its plunger has a soft brush attached to its surface. Syringe B is attached close to the end of syringe A. The barrel of A is pushed against the surface of the substrate to be sampled; water is then poured into its barrel and its plunger pushed down and rotated so that the brush removes the periphyton from the substrate. The plunger of syringe B is then pulled back, thereby withdrawing the fluid plus periphyton into its own barrel, and after this syringe A can be removed from the substrate. After collection, samples, may require further concentration, e.g. by centrigugation, although the method needs to be chosen with care if the more delicate species of algae are to be preserved unharmed. The samples can be observed directly using an ordinary stage microscope, although phase contrast illumination improves the amount of detail one can see. If one is only interested in the diatoms and there is a lot of contaminating organic matter present, the samples may be air dried and subjected to acid digestion (see below), although most ordinary algae do not have silicaceous cell walls and would be destroyed.

Diatoms are normally extracted from tissues or similar solid substrates, such as soil, by heating with concentrated nitric acid for up to 48 hours (Pollanen, 1998; Hurlimann *et al.*, 2000). The digest is then centrifuged, the supernatent discarded and the pellet washed by one or two cycles of suspension in distilled water and centrigugation, after which the final pellet is observed using a stage microscope. This method relies on the silicaceous frustules remaining after all organic matter has been dissolved away – but some workers prefer to use less dangerous reagents and have adopted techniques based on DNA extraction (Kobayashi *et al.*, 1993) and ultrasonic digestion (Matsumoto and Fukui, 1993). The possibility of using PCR technology to identify the presence of diatoms has been attempted (Abe *et al.*, 2003) but, although the technique is sensitive, it cannot yet be used to identify sufficient species of diatoms to distinguish the locality at which a person drowned. The possibility of undertaking molecular algal profiling in a similar manner to the microbial profiling mentioned in Chapter 4 does not appear to have been investigated as a forensic tool.

Collecting and preserving testate amoebae for forensic analysis

Preparation consists of boiling the sample in distilled water for 10 minutes and then filtering through a 425 μm mesh to remove large particles of debris. The filtrate is then centrifuged and the supernatent discarded. The pellet can then be resuspended in a glycerol–water mix, mounted on glass microscope slides and then examined using a stage microscope (Davis and Wilkinson, 2004). There is no single taxonomic key that covers all the UK testate amoebae but Corbet (1973) and Charman *et al.* (2000) are both useful guides. Once identified, the data can be analysed using Tilia and TiliaGraph software packages in a similar way to pollen (see below).

Collecting and preserving plant material for forensic analysis

Fresh leaves, mushrooms or other vegetable matter would rapidly rot if placed in the sealed plastic bags normally used for forensic samples and should therefore be put into individual brown paper bags that are then labelled to identify the person collecting the evidence and given an exhibit number. Fleshy fruits should be preserved in a fluid fixative, such as 70 per cent w/v alcohol, whilst dry fruits and seeds may be kept in small paper envelopes. Leaves may be preserved by drying and pressing in a plant press before being attached to herbarium sheets for protection and storage. Alternatively, the material can be stored in a fluid fixative, as described above, although this usually leads to a loss of colour as the pigments are bleached from the sample. However, it does reduce the risk of insect and fungal damage and of contamination during storage. Faeces and gut contents should be collected into marked wide-mouthed containers and, ideally, analysed for plant material soon after collection. It can also be helpful to preserve some of the samples in 10 per cent v/v formol–saline should further analyses be required. If the samples are extremely fluid, they should be centrifuged beforehand to concentrate the material.

Botanical specimens are normally identified by traditional taxonomic techniques based on morphological features. Unfortunately, some plant species are notoriously difficult to distinguish from one another and forensic samples may consist of plant fragments that do not provide sufficient features for a reliable identification to species level. Consequently, there is increasing interest in the use of molecular techniques for plant identification. However, the techniques used successfully in molecular animal taxonomy cannot always be employed with plants. For example, the Consortium for the Barcoding of Life is currently overseeing three worldwide DNA barcoding projects aimed at providing a database of all the world's fish and bird species. Ultimately, the number of projects will expand to include other animal groups and also the plants. In animals, the sequence of the gene coding for mitochondrial cytochrome c oxidase subunit 1 (COI) has been identified as possessing the necessary combination of conservation and divergence to facilitate discrimination between species (Herbert et al., 2003) but in higher plants this gene exhibits a comparatively slow rate of evolution that limits its suitability. Kress et al. (2005) have therefore suggested two alternative regions: the nuclear internal transcribed spacer region and the plastid trnH-psbA intergenic spacer region. Using these two regions they identified sequences that could be used to barcode 99 species of plants that encompassed 80 genera and 53 families. Similarly, Ward et al. (2005) have developed a series of PCR-based assays for the identification of grasses.

Because pollen and other palynomorphs are so easily transferred, contamination of forensic samples needs to be rigorously guarded against. For example, samples for pollen analysis should be taken at the crime scene before other investigators disturb any surrounding vegetation or, in the case of rooms, any windows are opened or excessive air currents are caused by lots of people walking around and opening doors. Similarly, if the case involves the death of someone, relatives and friends should not be permitted to leave floral tributes nearby until the specimens have been collected. Extreme care should also be taken in the laboratory to avoid cross-contamination and analysis should ideally take place in a semi-sealed room with all windows and doors closed and the sample containers should be opened and processed within a laminar flow cabinet. A thorough guide to sampling procedures is outlined at http://www.bahid.org/docs/NCF_Env%20Prof.html. Pollen may be recovered from both within and on the outer surfaces of a body, from clothing and from objects such as spades, furniture and vehicles, but on each of these there are certain areas that are most likely to yield useful information. For example, from within a body, pollen should be extracted from the nasal cavities and the stomach, whilst on the outer body surfaces the hair and underneath the fingernails are good sources. Hairsprays and oils can make the hair sticky and help trap pollen grains. On clothing and shoes, any muddy patches are obvious places to sample. However, even if these cannot be seen by the naked eye, it is still possible to extract pollen and other palynomorphs, bearing in mind the nature of the case being investigated. For example, the location of palynomorph evidence on an aggressor and their victim may be different and indicate how the assault progressed. In cases of female sexual assault that occur out of doors, the male assailant is likely to have evidence on the elbows, forearms and knees of his clothing and on the front portion of his shoes. His victim is likely to have a matching pattern of evidence on the back of her clothing and the heels of her shoes or feet. The wheel arches, air filters, seats and foot mats are good places to search for evidence in vehicles. Horrocks (2004) has produced an extremely useful guide to the preparation of forensic samples for pollen analysis. The procedure varies with the nature of the specimen, e.g. whether it is a soil sample or an item of clothing. However, it usually consists of initial treatment with hot 10 per cent w/v potassium hydroxide to dissolve any humic material and generally break up the sample matrix – this process is called deflocculation. In the case of soil samples, this is usually followed by treatment with hydrochloric acid, to remove carbonates, and then hydrofluoric acid, to remove silica. Cellulose is removed by treating the samples with glacial acetic acid in a process some authors refer to as acetylation and others as acetolysis, whilst lignin and other organic material are removed using a bleaching agent (Horrocks, 2004). After processing is completed the pollen is stained using 0.1 per cent basic fuschin and mounted in glycerol jelly onto glass microscope slides. As can be seen, the whole process involves several dangerous chemicals, especially hydrofluoric acid, and therefore requires a dedicated laboratory and a great deal of care. It also requires access to a reference pollen collection for comparing specimens. After counting the numbers of each type of pollen grain or the spores in the case of fungi, ferns and mosses, the data are usually analysed using a computer software package such as Tilia and TiliaGraph (Grimm, 1981) to produce a pollen (or spore) diagram in which species can be grouped by family and arranged alphabetically within these groups.

Collecting invertebrates for forensic analysis

In a homicide investigation, representative samples of all invertebrates found on and around the body should be collected directly into separate containers (Figure 9.1). Collection sites should include the soil underneath and surround-

Figure 9.1 Biological samples should be collected from around the body. Mature blowfly larvae and pupae may occur 3–6 metres away although in this case the body is fresh and therefore there would not have been time for them to have completed their development

ing a body or object at a crime scene. If the body is at an advanced stage of decay and mature blowfly larvae are present, there is an increased need to look for pupae in the surrounding soil. This may require sampling up to 3–6 metres surrounding the body because some maggot species travel considerable distances when they are ready to pupate. Maggots move away from the light so they are most likely to be found in the darker regions. They also orientate themselves to vertical surfaces, so they are also likely to be found close to walls and similar obstacles that they have met whilst crawling across the surface. In addition to larvae and pupae, any eggs present should also be collected. Insects found on the surrounding vegetation or elsewhere in a room can be collected using a sweep net or knocked into pots as appropriate. Forceps and a fine brush should be used carefully to avoid damaging the specimens.

Records should be kept of the exact position of all invertebrates found at a crime scene and their relative abundance. Specimens should be kept separate. For example, maggots collected from the mouth should not be mixed with those collected from the ears or anywhere else on the body. If the body is lying on the ground, known volumes of soil should be hand-sorted to locate pupae and also extracted using a standard technique, such as Tullgren and Berlese funnels, to determine the presence of active invertebrates (e.g. mites, fly larvae, earthworms) (Jackson and Raw, 1966). Samples of maggots should include all developmental stages from the smallest to the largest, and any pupae that are present. Maggots should ideally be killed at the time of collection and a note made of the exact moment this was done. If they are killed later, a record should be made of the temperature they experience between the times of collection and the time they are killed. Sub-samples of all maggots should be reared to adulthood to confirm identification. Larvae that are not killed immediately after they are collected should be provided with food to ensure that their developmental rate is not affected by a period of starvation or their size is affected by desiccation. There are reports of maggots squeezing between the zips of body bags (Greenberg and Kunich, 2002). Consequently, if samples were not taken immediately, or precautions taken, it could be argued that the largest (and oldest) maggots might have escaped before the body reached the morgue or contamination has occurred through maggots moving between corpses. Another reason for taking samples immediately is that dead bodies may be treated with preservative when they reach the morgue and this can cause the larvae to shrink.

Crawling invertebrates associated with cases of neglect, food contamination and other forensic investigations should be collected in the same manner as that outlined above.

It is difficult and sometimes impossible to identify the larval stages of many Diptera. Consequently, where possible, a representative sample of the maggots should be reared to adulthood. Plastic plant propagators kept inside incubators at a known constant temperature make suitable rearing containers (Figure 9.2). Care must be taken to ensure that the maggots do not either des-

Figure 9.2 Blowfly larvae may be reared to maturity inside plastic cloches such as this. The cloche should be stored inside an incubator or a temperature recording device included in the cloche if the development rate of the maggots needs to be recorded

iccate or drown in excessive moisture. Maggots may be reared on liver or minced meat: enclosing the meat in loosely wrapped foil helps to stop both the meat and the maggots becoming desiccated. Sherman and Tran (1995) have described a simple artificial diet composed of a sterile 1:1 mixture of 3 per cent Bacto agar and pureed liver. This medium can be stored for long periods at room temperature, which ensures replicable conditions and reduces the smell. Blowfly maggots develop at different rates on different tissues (Kaneshrajah and Turner, 2004) and the species and age of the animal from which the tissues were derived may also affect development rate. Consequently, the diet that laboratory-reared maggots receive and the region of a dead body that maggots are collected from could both influence minimum time since death estimations. The rearing medium should be surrounded by a layer of dry, friable material (e.g. vermiculite) into which the maggots can move when they are ready to pupate.

Flying insects are collected using a butterfly net with a fine mesh. The net is swept back and forth, rotating the bag through 180° at each pass. The insects are then transferred to individual tubes after collection. The insects should be killed immediately after collection to prevent them from damaging themselves. The species composition of flies visiting corpses is not necessarily an indication of its state of decomposition. For example, although adult female blowflies are most attracted to fresh corpses to lay their eggs, they also visit corpses at a late stage of decomposition to obtain a protein meal, or, in the case of males, in search of mates.

Table 9.1 Summary of killing and preservation methods for soft- and hard-bodied invertebrates

	Soft-bodied invertebrates	Hard-bodied invertebrates
Killing methods	Near-boiling water KAA fluid	Ethyl acetate vapour Hydrogen cyanide Carbon dioxide Near-boiling water (not winged insects)
Preservation methods	Acetic alcohol 70% v/v alcohol	Pinning Carding (beetles) Acetic alcohol 70% v/v alcohol

Killing and preserving techniques for invertebrates
(Table 9.1)

Invertebrates that need to be preserved for further study or as evidence should always be killed quickly and humanely. The ideal killing bottle should have a wide mouth, to facilitate entry and egress, and have a secure lid. It is best to have several killing bottles to reduce the risk of mixing up samples and to speed up the collection procedure. Plastic bottles are lighter and less prone to breakage than glass. There are a variety of killing methods (see below) but whichever one is chosen a record should be made of how, where and when it was carried out in case there are disputes about whether the process might have affected the results.

Killing methods for hard-bodied invertebrates

Ethyl acetate is a highly effective killing agent against all invertebrates but can damage some plastics, so care needs to be taken when choosing a killing bottle. Liquid ethyl acetate should not be allowed to come into contact with the insect so first a layer of plaster of Paris is used to coat the base of the killing bottle. Once the plaster has dried, ethyl acetate is added to the plaster of Paris. The prepared killing bottle lasts for several hours/days before it needs to be replenished. Insects should be removed from an ethyl acetate killing bottle as soon as they are dead.

Chopped cherry laurel leaves are rapidly lethal for adult Diptera and Hymenoptera but generally less effective against Coleoptera and cockroaches. Cherry laurel (*Prunus laurocerasus*) leaves give off cyanide when cut or bruised, and a layer of chopped leaves at the base of a small container makes a highly effective killing bottle. The insects should be separated from the leaf fragments by a layer of tissue to make their retrieval easier. The leaves create a high humidity inside the container, which keeps the specimens in a relaxed state, but this creates problems for specimens such as Lepidoptera in which the wings may become 'waterlogged' and if specimens are left in for too long they can become mouldy. Laurel leaves may be kept whole for over a week before they are cut up to create a killing bottle.

Carbon dioxide gas makes a good killing agent but this is really only suitable for invertebrates that have been transported to the laboratory – gas cylinders are a bit heavy to take into field situations. Some insects are more susceptible than others. Blowfly adults are killed quickly but cockroaches (as always) are more resistant. Similarly, placing specimens in the freezer at –20°C is effective for adult hard-bodied insects. The temperature of the freezing compartment of a typical home fridge is seldom low enough to kill some insects and even after 24 hours they will revive and crawl away. It is not a suitable way to kill maggots because it may influence subsequent length measurements.

Killing methods for soft-bodied invertebrates

Near-boiling water provides an extremely quick and effective end for all invertebrates (Figure 9.3). It is the best means of killing soft-bodied invertebrates such as maggots and other insect larvae but it is not suitable for winged insects because the wings become waterlogged and it becomes difficult to set them afterwards. Specimens killed in this way should be dehydrated and preserved as soon as they are dead otherwise they tend to swell. Maggots that are left in water after they have died will also melanize very quickly, i.e. they will darken,

Figure 9.3 These maggots were initially the same size. One was first killed in near-boiling water whilst the other, contracted, maggot was placed directly into alcohol

and this can obscure some morphological features. Boiling water should not be used because it can cause the specimen to rupture and the internal organs will spurt out. Heating up water is not always a feasible option in field conditions but a couple of large thermos flasks can often supply sufficient water.

Maggots and other larval insects may be placed directly into tubes of the preservative KAA, which consists of the following ingredients: 95 per cent ethanol (80–100 ml), glacial acetic acid (20 ml), and kerosene (i.e. paraffin) (10 ml).

Preservation of hard-bodied invertebrates

Large Diptera, such as blowflies, and most other insects are normally mounted on entomological pins placed through the thorax. Small insects are often staged by pinning them to a Styrofoam 'stage' using small headless entomological pins. A large pin is then inserted through the stage and the label is attached to this pin. Male genitalia are often an important aid to identification. These may be extended using a fine pin, although it may be necessary to mount them separately or preserve them as a microscope slide preparation. Very small Diptera, such as Phoridae, and other small invertebrates may be stored in 70 per cent alcohol. Woodlice, centipedes and millipedes are also best stored in 70 per cent alcohol. Coleoptera (beetles) are traditionally glued to pieces of card using a water-based adhesive to facilitate removal. However, this method makes the observation of ventral features impossible. Coleoptera should therefore be either pinned through one of the elytra and mounted like Diptera or stored in alcohol.

Preservation of soft-bodied invertebrates

Soft-bodied invertebrates, such as maggots, should be killed before placing them in preservative, otherwise they will shrink and their morphological features become obscured. A typical procedure would be to kill the specimens by placing them into near-boiling water until they are dead and then dehydrate them through increasing alcohol concentrations (30–80 per cent) before storing them in acetic alcohol (3 parts 80 per cent ethanol and 1 part glacial acetic acid). Specimens preserved in alcohol alone tend to become brittle but the addition of acetic acid helps to keep them soft. Regardless of the storage medium, specimens may shrink with time and this possibility should be borne in mind if reviewing evidence from cases that are several years old.

Preservation of invertebrate eggs

Invertebrate eggs are difficult to identify to species and, where possible, a proportion of them should be allowed to hatch and the young reared to adulthood.

Blowfly eggs may be dehydrated through increasing alcohol concentrations (15–80 per cent) before storing them in acetic alcohol or placed directly into KAA.

Invertebrate identification techniques

Traditional taxonomists are a threatened species within UK academia and those who remain are an ageing population. The major museums continue to support their activities although funding continues to be a problem. However, their efforts have built upon the enormous amount of work done by our Victorian ancestors and the continuing activities of an army of gifted amateurs to ensure that the UK has one of the best described invertebrate faunas in the world. Identification keys are available for most groups but, unfortunately, many of these have been written for experienced professionals and can be extremely difficult to decipher. The Royal Entomological Society publishes a series of handbooks covering virtually all the insect orders in the UK whilst the publisher E.J. Brill produces the Fauna Entomologica Scandinavica series that also includes many UK insect species (e.g. Rognes, 1991). More 'user-friendly' taxonomic keys covering insects of forensic importance can be found in Erzinclioglu (1996) (simple keys to some blowfly adults and larvae) and Smith (1986) (excellent comprehensive keys to both adult and larval Diptera). Taxonomic keys for invertebrates other than insects tend to be scattered throughout the literature and can be hard to track down.

In many species of maggot, the morphology of the anterior and posterior spiracles and the cephaloskeleton are important taxonomic features and are also useful in determining the instar. The anterior and posterior spiracles can often be seen without any prior treatment but to observe fine detail it is usually necessary to prepare slide mounts. Small larvae may be left whole but by the time they reach the third instar it is usually better to cut them into pieces. Cut off either the posterior or anterior of the maggot as appropriate and boil it in 10 per cent w/v sodium hydroxide to remove the internal soft tissues. This is best done on a hot plate. Wash the specimen in several changes of distilled water and dehydrate through alcohol steps (30 per cent–absolute alcohol, 10 minutes in each), then clear in xylene (or similar clearing agent) and mount on a microscope slide in Histoclear (or similar mountant). Remember to place a cover slip on top of the specimen! If the specimen does not have to be preserved as evidence, it is not necessary to dehydrate it and it can be mounted directly in water after being boiled in sodium hydroxide.

The identification of invertebrate eggs is often accomplished using scanning electron microscopy. According to Sukontason *et al.* (2004), a simpler routine is to stain the eggs in 1 per cent w/v potassium hydroxide solution for 1 minute followed by dehydration in 15 per cent, 70 per cent, 95 per cent and absolute alcohol (1 minute in each) and then permanent mounting on a glass microscope slide. They state that the staining reveals sufficient morphological detail to enable identification of several forensically important species of Diptera.

Owing to the lack of traditionally trained taxonomists, many scientists, and not just forensic biologists, are investigating the use of DNA markers to identify species (e.g. Harvey *et al.*, 2003; Chen *et al.*, 2004; Zehner *et al.*, 2004b). This is an especially attractive prospect where maggots are concerned. Theoretically, one could measure the length of a maggot and then divide it into three portions. The anterior and posterior regions would be processed and preserved on microscope slides to determine the maggot's instar and enable later confirmation of identification by traditional means, whilst the central region would be processed to extract the DNA. Taxonomic keys are not available for all insect larvae of forensic importance because rearing them to adulthood can be time consuming and success is not guaranteed. By contrast, because there are standard protocols for DNA extraction and analysis, these procedures could be done by a trained laboratory technician. However, although human DNA profiling is a central feature of many homicide and sexual assault investigations, the use of molecular taxonomy for the identification of blowflies and other invertebrates of forensic importance remains at the experimental stage. This is partly a consequence of concerns over the reliability of the gene sequence data used in the identification process. Although scientists have isolated species-specific gene sequences for some blowflies and fleshflies, their effectiveness against a wide range of populations and their potential for cross-reactivity with a wide range of other species have usually not been tested. For example, are all populations identified by the sequence, are there problems caused by sub-species and species complexes, and is there no cross-reaction with gene sequences present in any of the other insects likely to be found in the same environment? These questions would be exploited by lawyers to cast doubt on the validity of the identification and hence the conclusions drawn from the evidence. Hence, until more information becomes available, molecular identification techniques will probably tend to be used more for confirmation of traditional morphological methods rather than on their own in forensic cases.

Collecting environmental information for forensic analysis

The following information relates to the collection of environmental data at a crime scene in cases of suspicious death (Table 9.2) although the nature of data required would be the same or similar for most other forensic investigations involving biological material.

The date and time that the body was discovered and the samples were collected and, in the case of living invertebrate material, the time it was killed should be recorded. The date affects the abundance and diversity of the invertebrate fauna and the plant flora, whereas recording the various times is important when calculating the minimum time since death using invertebrate evidence.

Table 9.2 Summary of information to be collected at the site of a homicide that would improve the accuracy of biological forensic evidence for determining when and where the victim died

Type of evidence	Records required
Temporal	Date and time of body's discovery Date and time maggots collected Date and time maggots killed
Climatic	Weather conditions Light Wind speed (or ventilation / draughts if indoors) Temperature of microclimate where body found (leave data logger to record for several days after body removed) Temperature of body Temperature of maggot feeding ball (if present) Records from nearest weather station
Botanical	Identification and distribution of surrounding vegetation
Biological	Identification of maggots and other invertebrates associated with corpse and its surroundings Presence of other corpses and their distance Presence of dead organic matter and its distance Nature of any material covering all or part of the body (e.g. clothing, wrappings)
Geological	Underlying and overlying (if present) soil or other substrates (e.g. concrete or brick)

The exact geographical location (map grid reference, GPS coordinates, height above sea level) should be recorded because this can affect both invertebrate and plant diversity and abundance. Exposure to sunlight, the accessibility of orifices and wounds and the presence or absence of clothes will all influence the ability of insects to locate and utilize the corpse. For example, even relatively shallow burial will prevent the corpse being colonized by many blowfly species. The nature and distribution of surrounding vegetation will influence the sorts of invertebrates and vertebrates that are present in the locality. This can affect their ability to locate the corpse and will also affect the distribution of pollen and other plant evidence. The physical and chemical nature of the soil or substrate, its moisture content, etc. should be recorded because this may affect the ability of soil organisms to colonize the body and the preservation of biological evidence. Similarly, maggots leaving a corpse in search of a pupation site will have to travel further if the body is lying on a hard, impermeable surface.

Unless a body has remained in a centrally heated house since the time the person died, it will have been exposed to temperature fluctuations. These must be taken into account when estimating the postmortem interval from insect development patterns. A record must therefore be made at the time the body is discovered. Ideally, a temperature data logger would be placed at the site to determine the temperature fluctuations over the following 24–48 hours, and longer if possible. These records can then be compared to those recorded at the nearest weather station (Archer, 2004). This is important because temperatures can differ significantly over the space of a few miles. If a body is discovered indoors, the temptation is always to open any windows and doors to reduce the smell, however this would result in a drop in temperature so all measurements should be made before the room is ventilated. Similarly, it would allow flying insects to enter and leave the room before the correct collection procedure could be instigated. The air temperature should be taken both at the crime scene and immediately beside the body because local fluctuations may be caused by protection from, or exposure to, wind currents and the sun. The temperature of the body should be recorded and compared to the air temperature in case corrections need to be made when calculating the minimum time since death. Similarly, if a 'maggot feeding ball' is present in the body, the temperature in the ball should be noted. Wind and rain can reduce the activity of some insects, even if the air temperature is suitable for their growth and development. For example, a body may not be colonized by blowflies as soon as might be expected.

The presence of other bodies in the locality, including those of small wild animals, such as birds or rats, and also organic waste, such as farmyard manure, should be recorded. If deprived of food, some maggot species will travel in search of more (Christopherson and Gibo, 1997) and, therefore, the possibility that maggots might have migrated from elsewhere needs to be considered. Where there is a source of rotting organic matter nearby, such as farmyard manure or a waste dump, the levels of fly activity (and that of other insects) may be higher than normal and a body might be colonized extremely rapidly; there is also a similar possibility of maggots moving onto the corpse from the surrounding region.

Quick quiz

1. State three health and safety issues to be considered when collecting biological material from a dead body.

2. If a dead body discovered in a lake in August is judged to have died from drowning 2 months previously, why might it be necessary to return in June the following year to sample the diatoms growing in the lake?

3. Why should fresh leaves and fungi not be placed in plastic sample bags?

4. Why should pollen sampling be one of the first tasks in a forensic investigation?

5. Giving reasons for your choices, state three regions on a dead body that should be sampled for pollen.

6. Why should maggots collected from different regions of a dead body be stored separately?

7. Giving reasons for your choices, state four records that should accompany any maggot samples when they are collected from a dead body.

8. How far away from a dead body should one look for blowfly pupae?

9. Why is it a good idea to rear sub-samples of maggots collected from a dead body through to adulthood?

10. Why should larval blowflies be killed in near-boiling water but adult blowflies killed using ethyl acetate vapour?

11. Why is it helpful to record environmental information at a murder site for 24–48 hours after the discovery of the body?

12. At a murder site, why is it important to note the proximity of dead bodies to one another, as well as in relation to dead wild animals or decaying organic matter?

Project work

Title

How does preservation affect the length of blowfly maggots?

Rationale

Maggot length is a crucial measurement when determining the age of a maggot for a minimum time since death calculation and is affected by many factors.

Method

It would be best to do this on a variety of species of blowfly maggot and at a variety of stages of their development because size and differences in cuticle thickness, etc. are likely to have an influence on the response. The maggots should be killed in hot water of varying temperatures and then left for varying times before they are measured. Other should be placed directly into preservatives and measured after they are dead or frozen for varying times and then allowed to thaw out before being measured. Those killed in hot water should be placed in preservatives after being measured and then re-measured at varying time intervals of up to several months.

Title

The extent to which blowfly maggots can escape from body bags and the effect of morgue preservation fluids upon them.

Rationale

Some sources suggest that maggots can escape from the zips of body bags – this could facilitate contamination between bodies or the loss of important specimens. It would be useful to know the extent to which this occurs. Bodies are sometimes treated with a preservation fluid when they reach the morgue and the effect of this on the maggots is uncertain. For example, do they die immediately, are they stimulated to rapidly crawl away (and hence try to escape the body bag) and does it affect the length of maggots?

Method

Body bags can be obtained from a commercial supplier. Known numbers of maggots of varying sizes can be placed within a body bag and provided with a source of food. The number and size of the maggots that squeeze through the zip can be recorded. The maggots can also be sprayed with typical morgue preservation fluids and the effect on their movement and length determined.

Title

How far will a blowfly maggot go to pupate?

Rationale

It is often stated that one should look for blowfly pupae within 3–6 metres of a corpse but it would be useful to be able to localize one's searches more effectively.

Method

Blowfly larvae that are ready to pupate are easily obtainable from an angling shop, or their departure from a naturally infested dead animal or meat bait may be observed. Video recording or time-lapse photography can be useful to allow long-term monitoring. The experiments may take place in both the laboratory and the field to determine the time of day they tend to set off from their food source, the direction they take and whether movement is random, how far they will travel before pupating, the effect of obstacles, the presence of other larvae that have already started to pupate, light, shade, etc.

Postscript: future directions in forensic biology

Forensic science is currently undergoing an exciting, rapidly evolving renaissance with many areas of research and development waiting to be explored. These will be led by studies on identification techniques using human DNA and these will then be picked up and adapted by other branches of biology. Particular promise is shown by DNA chip technology, which utilizes large arrays of DNA probes bound to a silicon, glass or polypropylene surface. Sometimes referred to as a 'Lab-on-a-chip', these would be the size of a credit card and be able to extract the DNA from a sample, amplify specific sequences and analyse them to provide a DNA profile (Wang, 2000; Fedrigo and Naylor, 2004). Theoretically, such devices could be used at the scene of a crime and the profile compared with those held on a national database to provide an instant identification. The expansion of such databases will be another key area of advancement. These technologies are, however, raising ethical and legal issues that must be addressed. Molecular biology is a complex subject and overburdened with abbreviations and abstruse terminology. Consequently, it can be hard for a jury to truly understand the strengths and weaknesses of the 'evidence' put before them. The same can also be said of many prosecution and defence lawyers who become blinded by science and seduced by statements such as there being a 'one in 16 billion chance of the DNA sample coming from another person' without questioning the nature of profiles or the collection and processing of the samples. There is therefore a need to develop a mechanism that would enable DNA-based evidence to be presented in court in a standardized and clear way so that both its strengths and weaknesses are apparent.

The existence of DNA databases presents dilemmas concerning who would have access to the material, whether it could be tampered with and whether all citizens should be considered potential criminals. Whilst such databases have led to the arrest of several rapists and murderers, it is also worth remembering that most of the victims of violence know their assailant and the crimes are a consequence of sudden, unplanned anger or opportunism and fuelled by drink or drugs or both. The highly intelligent, charismatic serial killer who is a staple feature of crime fiction and film plots is, mercifully, an exceedingly rare individual. Most violent crimes are committed by poorly educated people living in the less affluent parts of our society and their victims come from the same background. These crimes are usually solved by standard police procedures. Indeed,

Essential Forensic Biology, Alan Gunn
© 2006 John Wiley & Sons, Ltd

despite all the technical advances, pitifully few cases of rape and sexual assault result in successful prosecutions. The answer to many crimes does not, therefore, depend on ever more sophisticated technologies but on the appropriate use of existing ones coupled with effective police work.

There is a serious need to develop rapid, cheap identification systems suitable for domestic animals and livestock. For example, the illegal movement of livestock and fraudulent subsidy and compensation claims are known to be big problems in the European Union. The illegal movement of animals can present serious risks in spreading diseases, some of which can also infect humans. Furthermore, following natural disasters, such as flooding, animals can stray and in the absence of a means of permanent identification there is the opportunity for theft and confusion. Although both hot and cold branding are cheap and permanent, their effectiveness is often compromised by poor record-keeping and difficulties in transmitting information between interested parties. Many of the other systems presently in use suffer from being expensive, unreliable or laborious. DNA databases are likely to have limited effectiveness for the identification of individual animals because livestock and domestic animals are often inbred and therefore exhibit reduced genetic variability. Furthermore, the cost and speed of the molecular biological techniques would need to be greatly reduced to make their routine use financially acceptable. Nevertheless, a high degree of discrimination is not always required. For example, often one only needs to determine the presence of a species or breed within a food product or medicine. Animal and plant DNA databases are unlikely to raise the same ethical issues as human DNA databases, although they would require the same rigorous standards concerning the collection, analysis and storage of the information. Retinal identification systems have been trialled with racehorses but these have suffered from practical problems and iris scans may be more effective (Cordes, 2000). It would be interesting to evaluate the usefulness of these techniques for other high-value animals. Retinal and iris identification systems will, however, only be effective whilst the animals are alive.

Stable isotope analytical techniques hold great promise for determining the geographical origin of all sorts of biological material, from bones to illegal drugs. Once reliable databases become available, these techniques will undoubtedly be used more frequently. For example, bacteria are known to exhibit a stable isotope 'fingerprint' of their growth media and it may be possible to use this as a means of determining their source in the case of a bioterrorism incident (Horita and Vass, 2002). Similarly, stable isotopes could be used to establish whether an organism was collected from the wild rather than being raised in captivity or cultivated. This is particularly relevant for distinguishing between meat from game animals raised in captivity on game farms from that from animals that were poached in the wild. Similarly, in the exotic pet trade, it could differentiate between captive born and animals caught in the wild.

All of the current procedures for estimating time since death suffer from drawbacks that limit their effectiveness and reliability. The ideal method would be one that was not dependent on biological processes that occur after death because these are complicated by a wide variety of environmental and chemical factors. However, in the meantime more can be done to improve existing techniques. For example, the use of insects and other invertebrates in criminal cases is hampered by the lack of trained taxonomists and our knowledge of basic biology. Some of the taxonomic keys for invertebrates of forensic importance are not user-friendly and there is a need for simpler ones that are supported by clear illustrations. The forensic potential of many invertebrates, such as mites and nematodes, is unlikely to be exploited until such keys become available. The development of DNA technology to facilitate identification holds some promise, especially for the identification of maggots and other larval invertebrates for which taxonomic keys are difficult to use or do not exist. There is surprisingly little published information on the development rates of invertebrates of forensic importance in the UK (and even less for those in other parts of the world) and how these are affected by the environment, competition or genotype / phenotype. There is also a need for carefully conducted studies on the effects of larval density on development rate and how burning affects colonization and subsequent larval development. It would also be helpful if a standardized laboratory-rearing protocol could be agreed upon to enable an easier comparison between studies.

The roles of bacteria and viruses in human and veterinary medicine have been studied intensively for many years but they could also have considerable forensic potential. The DNA profiles of bacteria and viruses could be useful as indicators for personal identification, evidence of previous associations or movements, as geographical and environmental indicators in soil and water samples found on clothing, implements or vehicles and for determining the time since death.

In recent years, there have been several high-profile cases in the UK and America in which persons have been wrongly convicted owing to faulty forensic evidence. In order to maintain the credibility of forensic science, in all its branches, it is therefore essential to improve the collection, transport and analysis of forensic samples so that the most appropriate tests are undertaken and it becomes impossible for samples to become mixed up or contaminated. This will inevitably increase the cost and time it takes to undertake analyses. Furthermore, the vast range of analytical procedures that are now at our disposal will create further costs and time constraints. This is already having a knock-on effect on the legal system that is starting to concern civil liberty organizations. For example, in the UK, the proposal in Autumn 2005 to increase the period during which suspected terrorists can be held without trial was driven, at least in part, by the amount of time needed to undertake forensic examinations. Therefore, although accuracy and rigour are paramount in forensic testing, the requirements for procedures to be quick and affordable are also highly important considerations.

References

Abbas, A. and Rutty, G.N. (2005) Ear piercing affects earprints: the role of ear piercing in human identification. *Journal of Forensic Sciences* **50**, 386–392.

Abe, S., Suto, M., Nakamura, H., Gunji, H., Hiraiwa, K., Suzuki, T., Itoh, T., Kochi, H. and Hoshiai, G.I. (2003) A novel PCR method for identifying plankton in cases of death by drowning. *Medicine Science and the Law* **43**, 23–30.

Adams, B.J. (2003) Establishing personal identification based on specific patterns of missing, filled, and unrestored teeth. *Journal of Forensic Sciences* **48**, 487–496.

Adams, Z.J.O. and Hall, M.J.R. (2003) Methods used for the killing and preservation of blowfly larvae, and their effect on post-mortem larval length. *Forensic Science International* **138**, 50–61.

Allery, J-P., Telmon, N., Mieusset, R., Blanc, A. and Rouge, D. (2001) Cytological detection of spermatozoa: comparison of three staining methods. *Journal of Forensic Sciences* **46**, 349–351.

Alunni-Perret, V., Muller-Bolla, M., Laugier, J.P., Lupi-Pegurier, L., Bertrand, M.F., Staccini, P., Bolla, M. and Quatrehomme, G.R. (2005) Scanning electron microscope analysis of experimental bone hacking trauma. *Jornal of Forensic Sciences* **50**, 796–801.

Ambach, W., Ambach, E., Tributsch, W., Henn, R., and Unterdorfer, H. (1992) Corpses released from glacier ice – glaciological and forensic aspects. *Journal of Wilderness Medicine* **3**, 372–376.

Ames, C. and Turner, B. (2003) Low temperature episodes in development of blowflies: implications for postmortem interval estimation. *Medical and Veterinary Entomology* **17**, 178–186.

Amorim, A. and Pereira, L. (2005) Pros and cons in the use of SNPs in forensic kinship investigation: a comparative analysis with STRs. *Forensic Science International* **150**, 17–21.

Anderson, G.S. (1999) Wildlife forensic entomology: determining time of death in two illegally killed black bear cubs. *Journal of Forensic Sciences* **44**, 856–859.

Anon (2004a) Isotopic techniques to provide fresh evidence. *Chemistry and Industry* **19**, 14.

Anon (2004b) Live anthrax bacteria inadvertently sent to vaccine researchers. *Nature* **429**, 692.

Anon (2004c) A true test of leadership. *Nature* **430**, 277.

Applebaum, A. (2004) *Gulag: A History of the Soviet Camps.* London: Penguin Books.

Essential Forensic Biology, Alan Gunn
© 2006 John Wiley & Sons, Ltd

Archer, M.S. (2003) Annual variation and departure times of carrion insects at carcasses: implications for succession studies in forensic entomology. *Australian Journal of Zoology* **51**, 569–576.

Archer, M.S. (2004) The effect of time after body discovery on the accuracy of retrospective weather station ambient temperature corrections in forensic entomology. *Journal of Forensic Sciences* **49**, 553–559.

Archer, M.S. and Elgar, M.A. (1998) Cannibalism and delayed pupation in hide beetles, *Dermestes maculates* DeGeer (Coleoptera: Dermestidae) *Australian Journal of Entomology* **37**, 158–161.

Archer, M.S. and Elgar, A.A. (2003) Effects of decomposition on carcass attendance in a guild of carrion-breeding flies. *Medical and Veterinary Entomology* **17**, 263–271.

Archer, N.E., Charles, Y., Elliott, J.A. and Jickells, S. (2005) Changes in lipid composition of latent fingerprint residue with time after deposition on a surface. *Forensic Science International* (in press).

Arnaldos, M.I., Garcia, M.D., Romera, E., Pressa, J.J. and Luna, A. (2005) Estimation of postmortem interval in real cases based on experimentally obtained entomological evidence. *Forensic Science International* **149**, 57–65.

Asamura, H., Takayanagi, K., Ota, M., Kobayashi, K. and Fukushima, H. (2004) Unusual characteristic patterns of postmortem injuries. *Journal of Forensic Sciences* **49**, 1–3.

Avila, F.W. and Goff, M.L. (1998) Arthropod succession patterns onto burnt carrion in two contrasting habitats in the Hawaiian Islands. *Journal of Forensic Sciences* **43**, 581–586.

Bal, R. (2005) How to kill with a ballpoint: credibility in Dutch forensic science. *Science Technology and Human Values* **30**, 52–75.

Banasr, A., de la Grandmaison, G.L. and Durigon, M. (2003) Frequency of bone/cartilage lesions in stab and incised wounds fatalities. *Forensic Science International* **131**, 131–133.

Bass, B. and Jefferson, J. (2003) *Death's Acre*. London: Time Warner.

Beard, B.L. and Johnson, C.M. (2000) Strontium isotope composition of skeletal material can determine the birth place and geographical mobility of humans and animals. *Journal of Forensic Sciences* **45**, 1049–1061.

Bender, K., Schneider, P.M. and Rittner, C. (2000) Application of mtDNA sequence analysis in forensic casework for the identification of human remains. *Foresnic Science international* **113**, 103–107.

Benecke, M. (2001) A brief history of forensic entomology. *Forensic Science International* **120**, 2–14.

Benecke, M. and Lessig, R. (2001) Child neglect and forensic entomology. *Forensic Science International* **120**, 155–159.

Bereuter, T.L., Mikenda, W. and Reiter, C. (1997) Iceman's mummification – implications from infrared spectroscopical and histological studies. *Chemistry – a European Journal* **3**, 1032–1038.

Berkefeld, K. (1993) [A possibility for verifying condom use in sex offences] *Archiv fur kriminologie* **192**, 37–42. (in German)

Bidmos, M.A. (2005) On the non-equivalence of documented cadaver lengths to living stature estimates based on Fully's method on bones in the Raymond A. Dart Collection. *Journal of Forensic Sciences* **50**, 501–506.

Blackledge, R. (2005) Condom trace evidence: the overlooked traces. *Forensic Nurse*: www.forensicnursemag.com/articles/311feat6.html

Bock, H., Ronneberger, D.L. and Betz, P. (2000) Suicide in a lions' den. *International Journal of Legal Medicine* 114, 101–102.

Bogenhagen, D. and Clayton, D.A. (1974) The number of mitochondrial deoxyribonucleic acid genomes in mouse L and human HeLa cells. *Journal of Biological Chemistry* 249, 7991–7995.

Bonte, W. (1975) Tool marks in bones and cartilage. *Journal of Forensic Sciences* 20, 315–325.

Borgula, L.M., Robinson, F.G., Rahimi, M., Chew, K.E., Birchmaier, K.R., Owens, S.G., Kieser, J.A. and Tompkins, G.R. (2003) Isolation and genotypic comparison of oral streptococci from experimental bitemarks. *Journal of Forensic Odontostomatology* 21, 23–30.

Bourel, B., Hedouin, V., Martin-Bouyer, L., Becart, A., Tournel, G., Deveaux, M. and Gossett, D. (1999) Effects of morphine in decomposing bodies on the development of *Lucilia sericata* (Diptera: Calliphoridae). *Journal of Forensic Sciences* 44, 354–358.

Bourel, B., Callet, B., Hedouin, V. and Gosset, D. (2003) Flies eggs: a new method for the estimation of short term post-mortem interval. *Forensic Science International* 135, 27–34.

Bourel, B., Hedouin, V. and Gosset, D. (2004) Effects of various substances on the delay of colonisation by necrophagous insects. *Proceedings of the 2nd Meeting of the European Association for Forensic Entomology*, London, 29–30 March 2004.

Bourel, B., Hubert, N., Hedouin, V. and Gosset, D. (2000) Forensic entomology applied to a mummified corpse. *Annales de la Societe Entomologique de France* 36, 287–290.

Brace, C.L. (1995) Region does not mean 'race' – reality versus convention in forensic anthropology. *Journal of Forensic Sciences* 40, 171–175.

Brain, C.K. (1981) *The Hunters or the Hunted? An Introduction to African Cave Taphonomy*. Chicago: University of Chicago Press.

Brattstrom, B.H. (1998) Forensic herpetology II: the McMartin preschool child molestation case and alleged animal sacrifice and torture, including reptiles. *Bulletin of the Chicago Herpetological Society* 33, 253–257.

Brauner, P., Reshef, A. and Gorski, A. (2001) DNA profiling of trace evidence – mitigating evidence in a dog biting case. *Journal of Forensic Sciences* 46, 1232–1234.

Breeze, R.G., Budowle, B. and Schutzer, S.E. (2005) *Microbial Forensics*. Amsterdam: Elsevier.

Brookes, A.J. (1999) The essence of SNPs. *Gene* 234, 177–186.

Brumfiel, G. (2003) Still out in the cold. *Nature* 423, 678–680.

Buck, T.J. and Vidarsdottir, U.S. (2004) A proposed method for the identification of race in sub-adult skeletons: a geometric morphometric analysis of mandibular morphology. *Journal of Forensic Sciences* 49, 1159–1164.

Budowle, B., Wilson, M.R., DiZinno, J.A., Stauffer, C., Fasano, M.A., Holland, M.M. and Monson, K.L. (1999) Mitochondrial DNA regions HVI and HVII population data. *Forensic Science International* 103, 23–35.

Bunyard, B.A. (2005) Commentary on: Carter DO, Tibbett M. Taphonomic mycota: fungi with forensic potential. *Journal of Forensic Sciences* 49, 1134.

Butler, J.M., Shen, Y. and McCord, B.R. (2003) The development of reduced size STR amplicons as a tool for the analysis of degraded DNA. *Journal of Forensic Sciences* **48**, 1054–1064.

Byard, R.W. (2004) Unexpected infant death: lessons from the Sally Clark case. *Medical Journal of Australia* **181**, 52–54.

Byard, R.W., James, R.A. and Gilbert, J.D. (2002) Diagnostic problems associated with cadaveric trauma from animal activity. *American Journal of Forensic Medicine and Pathology* **23**, 238–244.

Byers, S.N. and Myster, S. (2005) *Forensic Anthropology Laboratory Manual*. Needham Heights, MA: Allyn and Bacon.

Byrd, J.H. and Allen, J.C. (2001) Computer modelling of insect growth and its application to forensic entomology. In: Byrd, J.H. and Castner, J.L. *Forensic Entomology*, pp. 303–329. Boca Raton, FL: CRC Press.

Byrd, J.H. and Castner, J.L. (2001) *Forensic Entomology*. Boca Raton, FL: CRC Press.

Campobasso, C.P, Di Vella, G. and Introna, F. (2001) Factors affecting decomposition and Diptera colonization. *Forensic Science International* **120**, 18–27.

Campobasso, C.P., Marchetti, D. and Introna, F. (2004) Post-mortem artefacts made by ants and the effects of ant activity on decomposition rates. *Proceedings of the 2nd Meeting of the European Association for Forensic Entomology*. London, 29–30 March 2004.

Campobasso, C.P., Linville, J.G., Wells, J.D. and Introna, F. (2005) Forensic genetic analysis of insect gut contents. *American Journal of Forensic Medicine and Pathology* **26**, 161–165.

Cardinetti, B., Ciampini, C., D'Onofrio, C., Orlando, G., Gravina, L., Ferrari, F., *et al.* (2004) X-ray mapping technique: a preliminary study in discriminating gunshot residue particles from aggregates of environmental occupational origin. *Forensic Science International* **143**, 1–19.

Carter, D.O. and Tibbett, M. (2003) Taphonomic mycota: fungi with forensic potential. *Journal of Forensic Sciences* **48**, 168–171.

Casamatta, D.A. and Verb, R.G. (2000) Algal colonization of submerged carcasses in a mid-order woodland stream. *Journal of Forensic Sciences* **45**, 1280–1285.

Catts, E.P. and Goff, M.L. (1992) Forensic entomology in criminal investigations. *Annual Review of Entomology* **37**, 253–272.

Chang, Y.M., Burgoyne, L.A. and Both, K. (2003) High failure of amelogenin test in an Indian population group. *Journal of Forensic Sciences* **48**, 1309–1313.

Chapenoire, S. and Benezech, M. (2003) Forensic Medicine in Bordeaux in the 16th century. *American Journal of Medical Pathology* **24**, 183–186.

Charman, D.J., Hendon, D. and Woodland, W. (2000) *The identification of testate amoebae (Protozoa, Rhizopoda) from British oligotrophic peats*. QRA Technical Guide No. 9. London: Quaternary Research Association.

Chen, W.Y., Hung, T.H. and Shiao, S.F. (2004) Molecular identification of forensically important blow fly species (Diptera: Calliphoridae) in Taiwan. *Journal of Medical Entomology* **41**, 47–57.

Chisum, J. (2000) A commentary on bloodstain analysis in the Sam Sheppard case. *Journal of Behavioral Profiling* **1** (on line www.law-forensic.com/bloodstain_2.htm).

Christopherson, C. and Gibo, D.L. (1997) Foraging by food deprived larvae of *Neobellieria bullata* (Diptera: Sarcophagidae). *Journal of Forensic Sciences* **42**, 71–73.

Ciesielski, C.A., Marianos, D.W., Schochetman, G., Witte, J.J. and Jaffe, H.W. (1994) The 1990 Florida dental investigation – the press and the science. *Annals of Internal Medicine* **121**, 886–888.

Cooper, J.E. (1998) What is forensic veterinary medicine? Its relevance to the modern exotic animal practice. *Seminars in Avian and Exotic Pet Medicine* **7**, 161–165.

Corbet, S.A. (1973) An illustrated introduction to the testate Rhizopods in *Sphagnum* with special reference to the area around Malham Tarn, Yorkshire. *Field Studies* **3**, 801–838.

Cordes, T. (2000) Equine identification: the state of the art. *American Association of Equine Practitioners. Annual Convention Proceedings* **46**, 300–301.

Correia, A. and Pina, C. (2000) Tubercle of Carabelli: a review. *Dental Anthropology Journal* **15**, 18–21.

Courtin, G.M. and Fairgrieve, S.I. (2004) Estimation of postmortem interval (PMI) as revealed through the analysis of annual growth rings in woody tissue. *Journal of Forensic Sciences* **49**, 1–3.

Cox, T.M., Jack, N., Lofthouse, S., Watling, J., Haines, J. and Warren, M.J. (2005) King George III and porphyria: an elemental hypothesis and investigation. *The Lancet* **366**, 332–335.

Coyle, H.M.C. (2004) *Foresic Botany*. Boca Raton, FL: CRC Press.

Crosby, T.K. and Watt, J.C. (1986) Entomological identification of the origin of imported cannabis. *Journal of Forensic Science* **26**, 35–44.

Dadds, M.R., Turner, C.M. and McAloon, J. (2002) Development links between cruelty to animals and human violence. *Australian and New Zealand Journal of Criminology* **35**, 363–382.

Dalton, R. (2005) Infection scare inflames fight against biodefence network. *Nature* **433**, 344.

D'Andrea, F., Fridez, F. and Coquoz, R. (1998) Preliminary experiments on the transfer of animal hair during simulated criminal behaviour. *Journal of Forensic Sciences* **43**, 1257–1258.

Daugman, J. (2004) How iris recognition works. *IEE Transactions on Circuits and Systems for Video Technology* **14**, 21–30.

Daugman, J. and Downing, C. (2001) Epigenetic randomness, complexity and singularity of human iris patterns. *Proceedings of the Royal Society of London. Series B* **268**, 1737–1740.

Davies, D., Graves, D.J., Landgren, A.J., Lawrence, C.H., Lipsett, J., MacGregor, D.P. and Sage, M.D. (2004) The decline of the hospital autopsy: a safety and quality issue for healthcare in Australia. *Medical Journal of Australia* **180**, 281–285.

Davis, S.R. and Wilkinson, D.M. (2004) The conservation management value of testate amoebae as 'restoration' indicators: speculations based on two damaged raised mires in northwest England. *The Holocene* **14**, 135–143.

De Boek, G., Wood, M. and Samyn, N. (2002) Recent applications of LC-MS in forensic science. *Recent Applications in Forensic Science* (www.icgeurope.com).

De Greef, S. and Willems, G. (2005) Three dimensional cranio-facial reconstruction in forensic identification: latest progress and new tendencies in the 21st century. *Journal of Forensic Sciences* **50**, 12–17.

Denic, N., Huyer, D.W., Sinal, S.H., Lantz, P.E., Smith, C.R. and Silver, M.M. (1997) Cockroach: The omnivorous scavanager – potential misinterpretation of postmortem injuries. *American Journal of Forensic Medicine and Pathology* **18**, 177–180.

Dent, B.B., Forbes, S.L. and Stuart, B.H. (2004) Review of human decomposition processes in soil. *Environmental Geology* **45**, 576–585.

DiMaio, V.J.M. (1999) *Gunshot Wounds: Practical Aspects of Firearms, Ballistics, and Forensic Techniques* (2nd edn). Boca Raton, FL: CRC Press.

DiMaio, V.J.M. and DiMaio, D.D. (2001) *Forensic Pathology* (2nd edn). Boca Raton, FL: CRC Press.

DiNunno, N., Constantinides, F., Cina, S.J., Rizzardi, C., DiNunno, C. and Mauro, M. (2002) What is the best sample for determining the early postmortem period by on-the-spot flow cytometry analysis? *American Journal of Forensic Medicine and Pathology* **23**, 173–180.

Dire, D.J. (1992) Emergency management of dog and cat bite wounds. *Emergency Medicine Clinics of North America* **10**, 719–736.

Disney, R.H.L. (1994) *Scuttle Flies: The Phoridae.* New York: Chapman & Hall.

Disney, R.H.L. and Munk, T. (2004) Potential use of Braconidae (Hymenoptera) in forensic cases. *Medical and Veterinary Entomology* **18**, 442–444.

Dix, J. (2000) *Time of Death, Decomposition and Identification: an Atlas.* Boca Raton, FL: CRC Press.

DiZinno, J.A., Lord, W.D., Collins-Morton, M.B., Wilson, M.R. and Goff, M.L. (2002) Mitochondrial DNA sequencing of beetle larvae (*Nitidulidae: Omosita*) recovered from human bone. *Journal of Forensic Sciences* **47**, 1337–1339.

Douceron, H., Deforges, L., Gherardi, R., Sobel, A. and Chariot, P. (1993) Long lasting postmortem viability of human immunodeficiency virus – a potential risk in forensic medicine. *Forensic Science International* **60**, 61–66.

Du Plessis, R., Webber, L. and Saayman, G. (1999) Bloodborne viruses in forensic medical practice in South Africa. *American Journal of Forensic Medicine and Pathology* **20**, 364–368.

Duric, M., Rakocevic, Z. and Donic, D. (2005) The reliability of sex determination of skeletons from forensic context in the Blakans. *Forensic Science International* **147**, 159–164.

Dyer, C. (2005) Pathologist in Sally Clark case suspended from court work. *British Medical Journal* **330**, 1347.

Edgar, H.J.H. (2005) Predicting race using characters of dental morphology. *Journal of Forensic Sciences* **50**, 269–273.

Edston, E. and van Hage-Hamsten, M. (2003) Death in anaphylaxis in a man with house dust mite allergy. *International Journal of Legal Medicine* **117**, 299–301.

Ehleringer, J.R., Casale, J.F., Lott, M.J. and Ford, V.L. (2000) Tracing cocaine with stable isotopes. *Nature* **408**, 311–312.

Eichmann, C., Berger, B., Reinhold, M., Lutz, M. and Parson, W. (2004) Canine-specific STR typing of saliva traces on dog bite wounds. *International Journal of Legal Medicine* **118**, 337–342.

Elliott, S., Lowe, P. and Symmonds, A. (2004) The possible influence of microorganisms and putrefaction in the production of GHB in post-mortem biological fluid. *Forensic Science International* **139**, 183–190.

Erdtman, G. (1969) *Handbook of Palynology.* New York: Hafner Publishing Co.

Erzinclioglu, Z. (1996) *Blowflies.* Naturlaists' Handbooks 23. Slough, UK: The Richmond Publishing Co. Ltd.

Erzinclioglu, Z. (2000) *Maggots, Murder and Men.* Colchester: Harley Books.

Evans, E.P. (1906) *The Criminal Prosecution and Capital Punishment of Animals.* London: Heinemann. (Reprinted by Faber & Faber, London, 1988.)

Evershed, R.P. (1992) Chemical composition of a bog body adipocere. *Archaeometry* **34**, 253–265.

Fedrigo, O. and Naylor, G. (2004) A gene specific DNA sequencing chip for exploring molecular evolutionary change. *Nucleic Acids Research* **32**, 1208–1213.

Fiedler, S. and Graw, M. (2003) Decomposition of buried corpses, with special reference to the formation of adipocere. *Naturwissenschaften* **90**, 291–300.

Fojtasek, L. and Kmjec, T. (2005) Time periods of GSR particles deposition after discharge-final results. *Forensic Science International* **153**, 132–135.

Forbes, S.L., Stuart, B.H., Dadour, R.I. and Dent, B.B. (2004) A preliminary investigation of the stages of adipocere formation. *Journal of Forensic Sciences* **49**, 1–9.

Forbes, S.L., Dent, B.B. and Stuart, B.H. (2005a) The effect of soil type on adipocere formation. *Forensic Science International* **154**, 35–43.

Forbes, S.L., Stuart, B.H. and Dent, B.B. (2005b) The effect of the burial environment on adipocere formation. *Forensic Science International* **154**, 35–43.

Forbes, S.L., Stuart, B.H. and Dent, B.B. (2005c) The effect of method of burial on adipocere formation. *Forensic Science International* **154**, 44–52.

Forbes, S.L., Stuart, B., Dent, B. and Fenwick-Mulcahy, S. (2005d) Characterization of adipocere formation in animal species. *Journal of Forensic Sciences* **50**, 633–640.

Forbes, T.R. (1985) *Surgeons at the Bailey.* New Haven, CT: Yale University Press.

Fridez, F., Rochat, S. and Coquoz, R. (1999) Individual identification of cats and dogs using mitochondrial DNA tandem repeats? *Science & Justice* **39**, 167–171.

Gaillard, Y., Krishnamoorthy, A. and Bevalot, F. (2004) *Cerbera odollam*: a 'suicide tree' and cause of death in the state of Kerala, India. *Journal of Ethnopharmacology* **95**, 123–126.

Gheradi, M and Constantini, G. (2004) Death, elderly neglect, and forensic entomology. *Proceedings of the 2nd Meeting of the European Association for Forensic Entomology*, London, 29–30 March 2004.

Galloway, A. and Snodgrass, J.J. (1998) Biological and chemical hazards of forensic skeletal analysis. *Journal of Forensic Sciences* **43**, 940–948.

Gibbons, A. (1998) Calibrating the mitochondrial clock. *Science* **279**, 28–29.

Gill, P. (2001) An assessment of the utility of single nucleotide polymorphisms (SNPs) for forensic purposes. *International Journal of Legal Medicine* **114**, 204–210.

Gill, P., Ivanov, P.L., Kimpton, C., Piercy, R., Benson, N., Tully, G., *et al.* (1994) Identification of the remains of the Romanov family by DNA analysis. *Nature Genetics* **6**, 130–135.

Gill, P., Werrett, D.J., Budowle, B. and Guerrieri, R. (2004) An assessment of whether SNPs will replace STRs in national DNA databases – joint considerations of the DNA working group of the European Network of Forensic Science Institutes (ENFSI) and the Scientific Working Group on DNA Analysis Methods (SWGDAM). *Science and Justice* **44**, 51–53.

Goff, M.L. (2000) *A Fly for the Prosecution: How Insect Evidence Helps Solve Crimes.* Cambridge, MA: Harvard University Press.

Goff, M.L. and Lord, W.D. (2001) Entomotoxicology: insects as toxicological indicators and the impact of drugs and toxins on insect development. In: Byrd, J.H. and Castner, J.L. (2000) *Forensic Entomology*, pp. 331–340. Boca Raton, FL: CRC Press.

Goff, M.L., and Win, B.H. (1997) Estimation of postmortem interval based on colony development time for *Anoplepsis longipes* (Hymenoptera: Formicidae). *Journal of Forensic Sciences* 42, 1176–1179.

Goff, M.L., Brown, W.A., Hewadikaram, K.A. and Omori, A.I. (1991) Effect of heroin in decomposing tissues on the development rate of *Boettcherisca peregrina* (Diptera: Sarcophagidae) and implications of this effect on estimations of postmortem intervals using arthropod development patterns. *Journal of Forensic Sciences* 36, 537–542.

Gosline, A. (2005) Sperm clock calls time on rape. *New Scientist* 185, 12.

Grassberger, M. and Frank, C. (2003) Temperature-related development of the parasitoid wasp *Nasonia vitripennis* as a forensic indicator. *Medical and Veterinary Entomology* 17, 257–262.

Grassberger, M. and Reiter, C. (2001) Effect of temperature on *Lucilia sericata* (Diptera: Calliphoridae) development with special reference to the isomegalen- and isomorphen-diagram. *Forensic Science International* 120, 32–36.

Grassberger, M., Friedrich, E. and Reiter, C. (2003) The blowfly *Chrysomyia albiceps* (Weidmann) (Diptera: Calliphoridae) as a new forensic indicator in Central Europe. *International Journal of Legal Medicine* 117, 75–81.

Green, J.L., Holmes, A.J., Westoby, M., Oliver, I., Briscoe, D., Dangerfield, M., Gillings, M. and Beattie, A.J. (2004) Spatial scaling of microbial eukaryote diversity. *Nature* 432, 747–750.

Green, P.W.C., Simmonds, M.S.J. and Blaney, W.M. (2003) Diet nutriment and rearing density affect the growth of the black blowfly larvae, *Phormia regina* (Diptera: Calliphoridae). *European Journal of Entomology* 100, 39–42.

Green, S.T. and Wilson, O.F. (1996) The effect of hair color on the incorporation of methadone into hair in rat. *Journal of Analytical Toxicology* 20, 121–123.

Greenbaum, A.R., Donne, J., Wilson, D. and Dunn, K.W. (2004) Intentional burn injury: an evidence-based, clinical and forensic review. *Burns* 30, 628–642.

Greenberg, B. and Kunich, J.C. (2002) *Entomology and the Law.* Cambridge: Cambridge University Press.

Greenfield, H.J. (1988) Bone consumption by pigs in a contemporary Serbian village: implications for the interpretation of a prehistoric faunal assemblage. *Journal of Field Archaeology* 15, 473–478.

Grimaldi, L., De Giorgio, F., Masullo, M., Zoccai, G.B., Martinotti, G. and Rainio, J. (2005) Suicide by pencil. *Journal of Forensic Sciences* 50, 913–914.

Grimm, E.C. (1981) *Tilia and TiliaGraph.* Springfield: University of California.

Haefner, J.N., Wallace, J.R. and Merritt, R.W. (2004) Pig decomposition in lotic aquatic systems: the potential use of algal growth in establishing a postmortem submersion interval. *Journal of Forensic Sciences* 49, 477–480.

Haglund, W.D. and Sperry, K. (1993) The use of hydrogen peroxide to visualize tattoos obscured by decomposition and mummification. *Journal of Forensic Sciences* 38, 147–150.

Haglund, W.D., Reay, D.T. and Swindler, D.R. (1988) Tooth mark artefacts and survival of bones in animal scavenged human skeletons. *Journal of Forensic Sciences* **33**, 985–997.

Harvey, M.L., Mansell, M.W., Villet, M.H. and Dadour, I.R. (2003) Molecular identification of some forensically important blowflies of southern Africa and Australia. *Medical and Veterinary Entomology* **17**, 363–369.

Hassett, A.L., Radvanski, D.C., Gandiga, P.C., Escobar, J.I., Buyske, S.G. and Sigal, L.H. (2003) Co-morbid mental disorders in chronic Lyme disease: the first 50 patients. *Arthritis and Rheumatism* **48**, 1283.

Hayes, E.J and Wall, R. (1999) Age-grading adult insects: a review of techniques. *Physiological Entomology* **24**, 1–10.

Hayes, E.J., Wall, R. and Smith, K.E. (1998) Measurement of age and population structure in the blowfly *Lucilia sericata* (Meigen) (Diptera: Calliphoridae). *Journal of Insect Physiology* **44**, 895–901.

Hedges, D.J., Walker, J.A., Callinan, P.A., Shewale, J.G., Sinha, S.K. and Batzer, M.A. (2003) Mobile element-based assay for human gender determination. *Analytical Biochemistry* **312**, 77–79.

Hedouin, V., Bourel, B., Becart, A., Tournel, G., Deveaux, M., Goff, M.L. and Gosset, D. (1999) Determination of drug levels in larvae of *Lucilia sericata* (Diptera: Calliphoridae) reared on rabbit carcasses containing morphine. *Journal of Forensic Sciences* **44**, 351–353.

Hellerich, U. (1992) [Tattoo pigment in regional lymph nodes – an identifying marker] *Archives Kriminologia* **190**, 163–170 (in German).

Herbert, P.D., Ratnasingham, S. and deWaard, J.R. (2003) Barcoding animal life: cytochrome *c* oxidase subunit 1 divergences among closely related species. *Proceedings in Biological Sciences* **270** Suppl 1: S96–99.

Hopkins D.W., Wiltshire, P.E.J. and Turner, B.D. (2000) Microbial characteristics of soils from graves: an investigation into the interface of soil microbiology and forensic science. *Applied Soil Ecology* **14**, 283–288.

Hopwood, A.J., Mannucci, A. and Sullivan, K.M. (1996) DNA typing from human faeces. *International Journal of Legal Medicine* **108**, 237–243.

Horita, J. and Vass, A.A. (2002) Stable isotope fingerprints of biological agents. *Journal of Forensic Sciences* **48**, 122–126.

Horner-Devine, M.C., Lage, M., Hughes, J.B. and Bohannan, B.J.M. (2004) A taxa–area relationship for bacteria. *Nature* **432**, 750–753.

Horrocks, M. (2004) Sub-sampling and preparing forensic samples for pollen analysis. *Journal of Forensic Sciences* **49**, 1–4.

Horrocks, M. and Walsh, K.A.J. (2001) Pollen on grass clippings: putting the suspect at the scene of the crime. *Journal of Forensic Sciences* **46**, 947–949.

Horrocks, M., Coulson, S.A. and Walsh, K.A.J. (1998) Forensic palynology: variation in the pollen content of soil surface samples. *Journal of Forensic Sciences* **43**, 320–323.

Horrocks, M., Coulson, S.A. and Walsh, K.A.J. (1999) Forensic palynology: variation in the pollen content of soil on shoes and on shoeprints in soil. *Journal of Forensic Sciences* **44**, 119–122.

Horsewell, J., Cordiner, S.J., Maas, E.W., Martin, T.M., Bjorn, K., Sutherland, W., *et al.* (2002) Forensic comparison of soils by bacterial community DNA profiling. *Journal of Forensic Sciences* **47**, 350–353.

Hsieh, H-M., Huang, L-H., Tsai, L-C., Kuo, Y-C., Meng, H-H., Linacre, A. and Lee, J.C-I. (2003) Species identification of rhinoceros horns using the cytochrome *b* gene. *Forensic Science International* **136**, 1–11.

Hurlimann, J., Feer, P., Elber, F., Niederberger, K., Dirnhofer, R. and Wyler, D. (2000) Diatom detection in the diagnosis of death by drowning. *International Journal of Legal Medicine* **114**, 6–14.

Huxley, A.K. and Finnegan, M. (2004) Human remains sold to the highest bidder! A snapshot of the buying and selling of human skeletal remains on eBay®, and Internet auction sites. *Journal of Forensic Sciences* **49**, 17–20.

Ikegaya, H. and Iwase, H. (2004) Trial of the geographical identification using JC viral genotyping in Japan. *Forensic Science International* **139**, 169–172.

Ikegaya, H., Iwase, H., Sugimoto, C. and Yogo, Y. (2002) JC virus genotyping offers a new means of tracing the origins of unidentified cadavers. *International Journal of Legal Medicine* **116**, 242–245.

Inman, K. and Rudin, N. (1997) *An Introduction to Forensic DNA Analysis*. Boca Raton, FL: CRC Press.

Iscan, M.Y., Loth, S.R. and Wright, R.K. (1984) Age estimation from the rib by phase estimation: white males. *Journal of Forensic Sciences* **29**, 1094–1104.

Ivanov, P.L., Wadhams, M.J., Roby, R.K., Holland, M.M., Weedn, V.W. and Parsons, T.J. (1996) Mitochondrial DNA sequence heteroplasmy in the Grand Duke of Russia Georgij Romanov establishes the authenticity of the remains of Tsar Nicholas II. *Nature Genetics* **12**, 417–420.

Jackson, R.M. and Raw, F. (1966) *Life in the Soil*. London: Edward Arnold.

Jaffe, H.W., McCurdy, J.M., Kalish, M.L., Liberti, T., Metellus, G., Bowman, B.H., *et al.* (1994) Lack of HIV transmission in the practice of a dentist with AIDS. *Annals of Internal Medicine* **121**, 855–865.

Jobling, M.A. and Gill, P. (2004) Encoded evidence: DNA in forensic analysis. *Nature Reviews: Genetics* **5**, 739–752.

Johnson, D.J., Martin, L.R. and Roberts, K.A. (2005) STR-typing of human DNA from human fecal matter using the QIAGEN QIAmp® Stool Mini Kit. *Journal of Forensic Sciences* **50**, 802–808.

Kahana, T., Almog, J., Levy, J., Schmelzer, E., Spier, Y. and Hiss, J. (1999) Marine taphonomy: adipocere formation in a series of bodies recovered from a single shipwreck. *Journal of Forensic Sciences* **44**, 897–901.

Kahler, K., Haber, J. and Seidel, H-P. (2003) Reanimating the dead: reconstruction of expressive faces from skull data. *ACM Transactions on Graphics* **22**, 554–561.

Kaneshrajah, G. and Turner, B. (2004) *Calliphora vicina* larvae grow at different rates on different body tissues. *International Journal of Legal Medicine* **118**, 242–244.

Keim, P., Smith, K.L., Keys, C., Takahashi, H., Kurata, T. and Kaufmann, A. (2001) Molecular investigation of the Aum Shinrikyo anthrax release in Kameido, Japan. *Journal of Clinical Microbiology* **39**, 4566–4577.

Kiel, W., Rolf, B. and Sachs, H. (2003) Evidence of condom residues. *Forensic Science International* **136**, 261.

Klotzbach, H., Krettek, R., Bratzke, H., Puschel, K., Zehner, R. and Amendt, J. (2004) The history of forensic entomology in German-speaking countries. *Forensic Science International* **144**, 259–263.

Knudsen, P.J. (1993) Cytology in ballistics. An experimental investigation of tissue fragments on full metal jacketed bullets using routine cytological techniques. *International Journal of Legal Medicine* **106**, 15–18.

Kobayashi, M., Yamada, Y., Zhang, W.D., Itakura, Y., Nagao, M. and Takatori, T. (1993) Novel detection of plankton from lung tissue by enzymatic digestion method. *Forensic Science International* **60**, 81–90.

Komar, D. and Beattie, O. (1998) Postmortem insect activity may mimic perimortem sexual assault clothing patterns. *Journal of Forensic Science* **43**, 792–796.

Kosa, F. and Castellana, C. (2005) New forensic anthropological appoachment for the age determination of human fetal skeletons on the base of morphometry of vertebral column. *Forensic Science International* **147**, S69–S74.

Kress, W.J., Wurdack, K.J., Zimmer, E.A., Weigt, L.A. and Janzen, D.H. (2005) Use of DNA barcodes to identify flowering plants. *Proceedings of the National Academy of Sciences of the United States of America* **102**, 8369–8374.

Kreutzer-Martin, H.W., Chesson, L.A., Lott, M.J., Dorigan, J.V. and Ehleringer, J.R. (2004) Stable isotope ratios as a tool in microbial forensics – Part 2. Isotopic variation among different growth media as a tool for sourcing origins of bacterial cells or spores. *Journal of Forensic Sciences* **49**, 961–967.

Krogman, W.M. and Iscan, M.Y. (1986) *The Human Skeleton in Forensic Medicine*. Springfield: Charles C. Thomas.

Kubiczek, P.A. and Mellen, P.F. (2004) Commentary on: Huxley, A.K. and Finnegan, M. (2004) Human remains sold to the highest bidder! A snapshot of the buying and selling of human skeletal remains on eBay®, and Internet auction sites. *Journal of Forensic Sciences* **49**, 1137.

Kücken, M. and Newell, A.C. (2004) A model for fingerprint formation. *Europhysics Letters* **68**, 141–146.

Kunos, C.A., Simpson, S.W., Russell, K.F. and Hershkovitz, I. (1999) First rib metamorphosis: its possible utility for human age-at-death estimation. *American Journal of Physical Anthropology* **110**, 303–323.

Kurlan, R. and Kaplan, E.L. (2004) The pediatric neuropsychiatric disorders associated with streptococcal infection (PANDAS) etiology for tics and obsessive-compulsive symptoms: hypothesis or entity? Practical considerations for the clinician. *Pediatrics* **113**, 883–886.

Laloup, M., Pien, K. and Wood, M., Samyn, N., Boonen, T., Grootaert, P., *et al.* (2003) The use of combined liquid chromatography-mass spectrometry method (LC-MS/MS) to detect nordiazepam and oxazepam in a single larva and puparium of *Calliphora vicina* (Diptera: Calliphoridae). *Forensic Science International* **136**, 316–317.

Lantz, P-G., Mattson, M., Wadstrom, T. and Radstrom, P. (1997) Removal of PCR inhibitors from human faecal samples through the use of an aqueous two-phase system for sample preparation prior to PCR. *Journal of Microbiological Methods* **28**, 159–167.

Lasseter, A.E., Jacobi, K.P., Farley, R. and Hensel, L. (2003) Cadaver dog and handler team capabilities in the recovery of buried human remains in the southeastern United States. *Journal of Forensic Sciences* **48**, 617–621.

Lazarus, H.M., Price, R.S. and Sorensen, J. (2001) Dangers of large exotic pets from foreign lands. *Journal of Trauma-Injury Infection and Critical Care* **51**, 1014–1015.

Leendertz, F.H., Ellerbrok, H., Baesch, C., Couacy-Hymann, E., Matz-Rensing, K., Hakenbeck, R., *et al.* (2004) Anthrax kills wild chimpanzees in a tropical rainforest. *Nature* **430**, 451–452.

Linch, C.A., Whiting, D.A. and Hoolland, M.M. (2001) Human hair histogenesis for the mitochondrial DNA forensic scientist. *Journal of Forensic Sciences* **46**, 844–853.

Linville, J.G. and Wells, J.D. (2002) Surface sterilization of a maggot using bleach does not interfere with mitochondrial DNA analysis of crop contents. *Journal of Forensic Sciences* **47**, 1055–1059.

Linville, J.G., Hayes, J. and Wells, J.D. (2004) Mitochondrial DNA and STR analyses of maggot crop contents: effects of specimen preservation technique. *Journal of Forensic Sciences* **49**, 341–344.

Lo, M-C., Lee, H-M., Lin, M-W. and Tzen, C-Y. (2005) Analysis of heteroplasmy in hypervariable region II of mitochondrial DNA in maternally related individuals. *Annals of the New York Academy of Sciences* **1042**, 130–135.

Lockwood, R. (2000) Animal cruelty and human violence: the veterinarian's role in making connection – the American experience. *Canadian Veterinary Journal – Revue Veterinaire Canadienne* **41**, 876–878.

Lord, W.D., Goff, M.L., Adkins, T.R. and Haskell, N.H. (1994) The black soldier fly *Hermetia illuscens* (Diptera: Stratiomyidae) as a potential measure of human post-mortem interval: observations and case histories. *Journal of Forensic Sciences* **39**, 215–222.

Lord, W.D., DiZinno, J.A., Wilson, M.R., Budowle, B., Taplin, M. and Meinking, T.L. (1998) Isolation, amplification, and sequencing of human mitochondrial DNA obtained from a human crab louse, *Pthirus pubis* (L.), blood meals. *Journal of Forensic Sciences* **43**, 1097–1100.

Lorenzini, R. (2005) DNA foresnics and the poaching of wildlife in Italy: a case study. *Forensic Science International* **153**, 218–221.

Love, J.C. and Symes, S.A. (2004) Understanding rib fracture patterns: incomplete and buckle fractures. *Journal of Forensic Sciences* **49**, 1153–1158.

Ludes, B., Coste, M., North, N., Doray, S., Tracqui, A. and Kintz, P. (1999) Diatom analysis in victim's tissues as an indicator of the site of drowning. *International Journal of Legal Medicine* **112**, 163–166.

Lunetta, P., Penttila, A. and Hallfors, G. (1998) Scanning and transmission electron microscopical evidence of the capacity of diatoms to penetrate the alveo- capillary barrier in drowning. *International Journal of Legal Medicine* **111**, 229–237.

Lunetta, P., Ohberg, A. and Sajantila, A. (2002) Suicide by intracerebellar ballpoint pen. *American Journal of Forensic Medicine and Pathology* **23**, 334–337.

Maher, J., Vintiner, S., Elliot, D. and Melia, L. (2002) Evaluation of the BioSign PSA membrane test for the identification of semen stains in forensic work. *New Zealand Medical Journal* **114**, 48–49.

Manfredi, G., Thyagarajan, D., Papadopoulou, L.C., Pallotti, F. and Schon, E.A. (1997) The fate of human sperm-derived mtDNA in somatic cells. *American Journal of Human Genetics* **61**, 953–960.

Marko, P., Lee, S.C., Rice, A.M., Gramling, J.M., Fitzhenry, T.M., McAlister, J.S., Harper, G.R. and Moran, A.L. (2004) Mislabelling of a depleted reef fish. *Nature* **430**, 309–310.

Matsuda, H., Seo, Y., Kakizaki, E., Kozawa, S., Muraoka, E. and Yukowa, N. (2004) Identification of DNA of human origin based on amplification of human-specific mitochondrial cytochrome b region. *Forensic Science International* **152**, 109–114.

Matsumoto, H. and Fukui, Y. (1993) A simple method for diatom detection in drowning. *Forensic Science International* **60**, 91–95.

Mejia, R. (2005) You can't rely on firearm forensics. *New Scientist* **188**, 6–7.

Mellen, P.F., Lowry, M.A. and Micozzi, M.S. (1993) Experimental observations on adipocere formation. *Journal of Forensic Sciences* **38**, 91–93.

Money, N.P. (2004) *Carpet Monsters and Killer Spores: A Natural History of Toxic Mold.* Oxford: Oxford University Press.

Monteiro, J.A. (1995) Human and animal bite wound infections. *European Journal of Internal Medicine.* **6**, 209–215.

Moran, N.C. and O'Connor, T.P. (1992) Bones that cats gnawed upon: a case study in bone modification. *Circaea* **9**, 27–34.

Mumcuoglu, K.Y., Gallili, K.Y., Reshef, A., Brauner, P. and Grant, H. (2004) Use of lice in forensic entomology. *Journal of Medical Entomology* **41**, 803–806.

Munro, H.M.C. and Thrushfield, M.V. (2001) 'Battered pets': sexual abuse. *Journal of Small Animal Practice* **42**, 333–337.

Munro, R. (1998) Forensic necroscopy. *Seminars in Avian and Exotic Pet Medicine* **7**, 201–209.

Murakami, H., Yamamoto, Y., Yoshitome, K., Ono, T., Okamoto, O., Shigeta, Y., *et al.* (2000) Forensic study of sex determination using PCR on teeth samples. *Acta Medica Okayama* **54**, 21–32.

Nakazono, T.O., Kashimura, S., Hayashiba, Y., Hara, K. and Miyoshi, A. (2005) Successful DNA typing of urine stains using a DNA purification kit following dialfiltration. *Journal of Forensic Sciences* **50**, 860–864.

Negrusz, A., Moore, C.M., Hinkel, K.B., Stokham, T.L., Verma, M., Strong, M.J. and Janicak, P.G. (2001) Deposition of 7-aminoflunitrezapam and flunitrezepam in hair after a single dose of Rohypnol®. *Journal of Forensic Sciences* **46**, 1143–1151.

Nields, H. and Kessler, S.C. (1998) Streptococcal toxic shock syndrome presenting as suspected child abuse. *American Journal of Forensic Medicine and Pathology* **19**, 93–97.

Nolte, K.B. and Yoon, S.S. (2003) Theoretical risk for occupational blood-borne infections in forensic pathologists. *Infection Control and Hospital Epidemiology* **24**, 772–773.

Nowak, R. (2004) Murder detectives must rethink maggot theory. *New Scientist* **3 April**, 13.

Nuorteva, P. (1977) Sarcosaprophagous insects as forensic indicators. In: Tedeschi, C.G. *et al.*, *Forensic Medicine*, Vol. 2, pp. 1072–1098. Philadelphia, PA: W.B. Saunders Co.

Nyhus, P.J., Tilson, R.L. and Tomlinson, J.L. (2003) Dangerous animals in captivity: *ex situ* tiger conflict and implications for private ownership of exotic animals. *Zoo Biology* **22**, 573–586.

Ohshima, T. (2000) Forensic wound examination. *Forensic Science International* **113**, 153–164.

Ohtani, S., Ito, R. and Yamamoto, T. (2003) Differences in the D/L aspartic acid ratios in dentin among different types of teeth from the same individual and estimated age. *International Journal of Legal Medicine* **117**, 149–152.

Oxley, J.C., Smith, J.L., Kirschenbaum, L.J., Shinde, K.P. and Marimganti, S. (2005) Accumulation of explosives in hair. *Journal of Forensic Sciences* 50, 826–831.

Ozdemir, M.H., Askoy, U., Akisu, C., Sonmez, E., and Cakmak, M.K. (2003) Investigating demodex in forensic autopsy cases. *Forensic Science International* 135, 226–231.

Pamplin, C. (2004) Science on trial. *Chemistry & Industry* 6 Dec., 14–15.

Peeters, M., Courgnaud, V., Abela, B., Auzel, P., Pourrut, X., Bibollet-Ruche, F., Loul, S., *et al.* (2002) Risk to human health from a plethora of Simian immunodeficiency viruses in primate bushmeat. *Emerging Infectious Diseases* 8, 451–457.

Peng, Z. and Pounder, D.J. (1998) Forensic medicine in China. *American Journal of Forensic Medicine and Pathology* 19, 368–371.

Peters, W. and Gilles, H.M. (1977) *A Colour Atlas of Tropical Medicine and Parasitology* (1st edn). London: Wolfe Medical Publications.

Pfeiffer, I., Volkel, I., Taubert, H. and Brenig, B. (2004) Forensic DNA typing of dog hair: DNA extraction and PCR amplification. *Forensic Science International* 141, 149–151.

Phillips, C., Lareu, V., Sala, A. and Carracedo, A. (2004) Nonbinary single-nucleotide polymorphism markers. *Progress in Forensic Genetics* 10, 27–29.

Pickering, T.R. (2001) Carnivore voiding: a taphonomic process with the potential for the deposition of forensic evidence. *Journal of Forensic Sciences* 46, 406–411.

Platek, S.F., Ranieri, N. and Wolnik, K.A. (1997) A false report of product tampering involving a rodent and soft drink can: light microscopy, image analysis and scanning electron microscopy/energy dispersive X-ray analysis. *Journal of Forensic Sciences* 42, 1171–1175.

Pollanen, M.S. (1997) The diagnostic value of the diatom test for drowning 2. Validity: Analysis of diatoms in bone marrow and drowning medium. *Journal of Forensic Sciences* 42, 286–290.

Pollanen, M.S. (1998) Diatoms and homicide. *Forensic Science International* 91, 29–34.

Pollanen, M.S., Cheung, L. and Chaisson, D.A. (1997) The diagnostic value of the diatom test for drowning 1. Utility: A retrospective analysis of 771 cases of drowning in Ontario, Canada. *Journal of Forensic Sciences* 42, 281–285.

Pounder, D.J. (1991) Forensic entomo-toxicology. *Journal of the Forensic Science Society* 31, 469–472.

Pretty, I. and Sweet, D. (2000) Anatomical location of bitemarks and associated findings in 101 cases from the United States. *Journal of Forensic Sciences* 45, 812–814.

Pretty, I. and Turnbull, M.D. (2001) Lack of dental uniqueness between two bite mark suspects. *Jounal of Forensic Sciences* 46, 1487–1491.

Prinz, M., Grellner, W. and Schmitt, C. (1993) DNA typing of urine samples following several years of storage. *International Journal of Legal Medicine* 106, 75–79.

Quarino, L., Dang, Q., Hartman, J. and Moynihan, N. (2005) An ELISA method for the identification of salivary amylase. *Journal of Forensic Sciences* 50, 873–876.

Quatrehomme, G. and Iscan, M.Y. (1997a) Postmortem skeletal lesions. *Forensic Science International* 89, 155–165.

Quatrehomme, G. and Iscan, M.Y. (1997b) Bevelling in exit gunshot wounds in bones. *Forensic Science International* 89, 93–101.

Rahimi, M., Heng, N.C.K., Kieser, J.A. and Tompkins, G.R. (2005) Genotypic copari-son of bacteria recovered from human bite marks and teeth using arbitrarily primed PCR. *Journal of Applied Microbiology* **99**, 1265–1270.

Raupp, C.D. (1999) Treasuring, trashing or terrorizing: adult outcomes of childhood socialization about companion animals. *Society & Animals* **7**, 141–159.

Rawson, R.B., Starich, G.H. and Rawson, R.D. (2000) Scanning electron microscope analysis of skin resolution as an aid in identifying trauma in forensic investigations. *Journal of Forensic Sciences* **45**, 1023–1027.

Ray, D.A., Walker, J.A., Hall, A., Llewellyn, B., Ballantyne, J., Christian, A.T., Turtle-taub, K. and Batzer, M.A. (2005) Inference of human geographic origins using *Alu* insertion polymorphisms. *Forensic Science International* **153**, 117–124.

Reid, M.E. and Lomas-Francis, C. (1996) *The Blood Group Antigen Facts Book*. San Diego, CA: Academic Press.

Roeterdink, E.M., Dadour, I.R. and Watling, R.J. (2004) Extraction of gunshot residues from the larvae of the forensically important blowfly *Calliphora dubia* (Macquart) (Diptera: Calliphoridae) *International Journal of Legal Medicine* **118**, 63–70.

Rognes, K. (1991) *Blowflies (Diptera, Calliphoridae) of Fennoscandia and Denmark*. Fauna Entomologica Scandinavica **24**. Leiden: E.J. Brill / Scandinavian Science Press Ltd.

Romain, N., Brand-Casadevall, C., Dimo-Simonin, N., Michaud, K., Mangin, P. and Papilloud, J. (2002) Post mortem castration by a dog: a case report. *Medicine and the Law* **42**, 269–271.

Rompen, J.C., Meek, M.F. and van Andel, M.S. (2000) A cause celebre: the so-called 'ballpoint murder'. *Journal of Forensic Sciences* **45**, 1144–1147.

Ropohl, D., Scheithauser, R. and Pollak, S. (1995) Postmortem injuries inflicted by domestic golden hamster: morphological aspects and evidence by DNA typing. *Forensic Science International* **72**, 81–90.

Rothschild, M.A. and Schneider, V. (1997) On the temporal onset of postmortem scav-enging 'motivation' of the animal. *Forensic Science International* **89**, 57–64.

Rudnik, I., Poplawska, R., Szulc, A., Juchnowicz, D., Konarzewska, B., Czernikiewicz, A. and Debowska, I. (2003) Mental disorders in patients with Lyme disease. *European Neuropsychopharmacology* **13**, S443.

Saukko, P. and Knight, B. (2003) *Knight's Forensic Pathology* (3rd edn). London: Arnold.

Savolainen, P. and Lundberg, J. (1999) Forensic evidence based on mtDNA from dog and wolf hairs. *Journal of Forensic Sciences* **44**, 77–81.

Schmitt, A. and Murail, P. (2004) Is the first rib a reliable indicator of age at death assessment? Test of the method developed by Kunos, *et al.* (1999). *Homo* **54**, 207–214.

Schneider, P.M., Seo, Y. and Rittner, C. (1999) Forensic mtDNA hair analysis excludes a dog from having caused a traffic accident. *International Journal of Legal Medicine* **112**, 315–316.

Schroeder, H., Klotzbach, H., Oesterhelweg, L. and Puschel, K. (2002) Larder beetles (Coleoptera, Dermestidae) as an accelerating factor for decomposition of a human corpse. *Forensic Science International* **127**, 231–236.

Schudel, D. (2001) Screening for canine spermatozoa. *Science & Justice* **41**, 117–119.

Schumm, J.W., Wingrove, R.S. and Douglas, E.K. (2004) Robust STR multiplexes for challenging casework samples. *Progress in Forensic Genetics* ICS **1261**, 547–549.

Scott, S. and Duncan, C. (2004) *Return of the Black Death: the World's Greatest Serial Killer.* Chichester: Wiley.

Scutt, J. (1990) Beware of new technologies. *Legal Service Bulletin* **15**, 9–12.

Selavka, C.M. (1991) Poppy seed ingestion as a contributing factor to opiate-positive urinalysis results: the Pacific perspective. *Journal of Forensic Sciences* **36**, 685–696.

Sharp, D. (1997) Infrared reveals the 'iceman's' adipocere.*The Lancet* **350**, 191.

Shepherd, R. (2003) *Simpson's Forensic Medicine* (12th edn). London: Arnold.

Sherman, R.A. and Tran, J.M.T. (1995) A simple, sterile food source for rearing the larvae of *Lucilia sericata* (Diptera, Calliphoridae). *Medical and Veterinary Entomology* **9**, 393–398.

Sidari, L., Di Nunno, N., Constantinides, F. and Melato, M. (1999) Diatom test with Soluene-350 to diagnose drowning in sea water. *Forensic Science International* **103**, 61–65.

Sims, R.W. and Gerard, B.M. (1985) *Earthworms*. Synopses of the British Fauna. 31. London: E.J. Brill/Dr. W. Backhuys Publishers.

Singh, R.R., and Goyal, S.P. (2003) Application of X-ray diffraction to characterize ivory, antler and rhino horn: implications for wildlife forensics. *Forensic Science International* **136**, 376–377.

Sinha, S.K., Budowle, B., Arcot, S.S., Richey, S.L., Chakraborty, R., Jones, M.D., Wojtkiewicz, P.W., *et al.* (2003) Development and validation of a multiplexed Y-chromosome STR genotyping system, Y-PLEX™6, for forensic casework. *Journal of Forensic Sciences* **48**, 93–103.

Skidmore, P. (1985) *The Biology of the Muscidae of the World*. The Hague: Dr W. Junk Publishers.

Skinner, M. and Dupras, T. (1993) Variation in birth timing and location of the neonatal line in human enamel. *Journal of Forensic Sciences* **38**, 1383–1390.

Smith, K.G.V. (1986) *A Manual of Forensic Entomology.* London: British Natural History Museum.

Spalding, K., Bhardwaj, R., Buchholz, B., Druid, H. and Frisen, J. (2005a) Retrospective birth dating of cells in humans. *Cell* **122**, 133–143.

Spalding, K., Buchholz, B.A., Bergman, L.E., Druid, H. and Frisen, J. (2005b) Age written in teeth by nuclear tests. *Nature* **437**, 333–334.

Sperry, K. (1991) Tattoos and tattooing. Part 1: History and methodology. *American Journal of Forensic Medicine and Pathology* **12**, 313–319.

Spiers, E.M. (1975) The use of the dum dum bullet in colonial warfare. *Journal of Imperial and Commonwealth History* **4**, 3–14.

Stanley, E.A. (1992) Application of palynology to establish the provenance and travel history of illicit drugs. *Microscope* **40**, 149–152.

Stelling, M.A. and van der Peijl, G.J.Q. (2003) Analytical chemical tools in wildlife forensics. *Forensic Science International* **136**, 381–382.

Stephen, C.N. and Henneberg, M. (2001) Building faces from dry skulls: are they recognized above chance rates? *Journal of Forensic Sciences* **46**, 432–430.

Stoermer, E.F. and Smol, J.P. (2001) *The Diatoms: Applications for the Earth and Environmental Sciences*. Cambridge: Cambridge University Press.

Stroud, R.K. (1998) Wildlife forensics and the veterinary practitioner. *Seminars in Avian and Exotic Pet Medicine* 7, 182–192.

Sukontason, K., Sukontason, K.L., Piangjai, S., Boonchu, N., Kurahashi, H., Hope, M. and Olson, J.K. (2004) Identification of forensically important fly eggs using potassium permanganate staining technique. *Micron* 35, 391–395.

Sung, Tz'u and McKnight, B.E. (1981) *The Washing Away of Wrongs: Forensic Medicine in Thirteenth Century China.* Ann Arbor, MI: University of Michigan Publishers.

Swift, B. and Rutty, G.N. (2003) The human ear: its role in forensic practice. *Journal of Forensic Sciences* 48, 153–160.

Swift, B., Lauder, I., Black, S. and Norris, J. (2001) An estimation of the post-mortem interval in human skeletal remains: a radionucleotide and trace element approach. *Forensic Science International* 117, 73–87.

Sykes, L.N., Champion, H.R. and Fouty, W.J. (1988) Dum-dums, hollow-points, and devastators – techniques designed to increase wounding potential of bullets. *Journal of Trauma-Injury Infection and Critical Care* 28, 618–623.

Taylor, J.J. (1994) Diatoms and drowning – a cautionary case note. *Medicine, Science, and the Law* 34, 78–79.

Thali, M.J., Braun, M., Markwalder, T.H., Brueschweiler, W., Zollinger, U., Malik, N.J., Yen, K. and Dirnhofer, R. (2003) Bite mark documentation and analysis: the forensic 3D/CAD supported photogrammetry approach. *Forensic Science International* 135, 115–121.

Tibbett, M., Carter, D.O., Haslam, T., Major, R. and Haslam, R. (2004) A laboratory incubation method for determining the rate of microbial degradation of skeletal muscle tissue in soil. *Journal of Forensic Sciences* 49, 1–6.

Tomberlin, J.K., Sheppard, D.C. and Joyce, J.A. (2005) Black soldier fly (Diptera: Stratiomyidae) colonization of pig carrion in south Georgia. *Journal of Forensic Sciences* 50, 152–153.

Tracqui, A., Fonmartin, K., Geraut, A., Pennera, D., Doray, S. and Ludes, B. (1998) Suicidal hanging resulting in complete decapitation: a case report. *International Journal of Legal Medicine* 112, 55–57.

Tracqui, A., Keyser-Tracqui, C., Kintz, P. and Ludes, B. (2004) Entomotoxicology for the forensic toxicologist: much ado about nothing? *International Journal of Legal Medicine* 118, 194–196.

Trotter, M. (1970) Estimation of stature from intact long limb bones. In: Stewart, T.D. (ed.). *Personal Identification in Mass Disasters.* Washington: Smithsonian Institution.

Tsokos, M. and Schulz, F. (1999) Indoor postmortem animal interference by carnivores and rodents: report of two cases and review of the literature. *International Journal of Legal Medicine* 112, 115–119.

Tsokos, M., Matschke, J., Gehl, A., Koops, E. and Puschel, K. (1999) Skin and soft tissue artefacts due to postmortem damage caused by rodents. *Forensic Science International* 104, 47–57.

Tun, Z., Honda, K., Nakatome, M., Nakamura, M., Shimada, S., Ogura, Y., *et al.* (1999) Simultaneous detection of multiple STR loci on sex chromosomes for forensic testing of sex and identity. *Journal of Forensic Sciences* 44, 772–777.

Vandenberg, N. and van Oorschot, R.A.H. (2002) Extraction of human nuclear DNA from feces samples using the QIAamp DNA Stool Mini Kit. *Journal of Forensic Sciences* **47**, 993–995.

Vanezis, P., Vanezis, M., McCombe, G. and Niblett, T. (2000) Facial reconstruction using 3-D computer graphics. *Forensic Science International* **108**, 81–95.

Varetto, L. and Curto, O. (2004) Long persistence of rigor mortis at constant low temperature. *Forensic Science International* **147**, 31–34.

Vasiliev, V., Vali, M. and Pallo, H. (2003) Forensic approach to fatal and non-fatal dog attacks: case reports. *Forensic Science International* **136**, 252–253.

Villain, A., Cheze, A., Tracqui, A., Ludes, B. and Kintz, P. (2004) Windows of detection of zolpidem in urine and hair: application of drug facilitated sexual assaults. *Forensic Science International* **143**, 157–161.

Vintiner, S.K., Stringer, P. and Kanagasundaram, S. (1992) Alleged sexual violation of a human female by a rottweiler dog. *Journal of the Forensic Science Society* **32**, 357–362.

Wang, J. (2000) From DNA biosensors to gene chips. *Nucleic Acids Research* **28**, 3011–3016.

Ward, J., Peakall, R., Gilmore, S.R. and Robertson, J. (2005) Molecular identification system for grasses: a novel technology for forensic botany. *Forensic Science International* **152**, 121–131.

Watkins, W.S., Rogers, A.R., Ostler, C.T., Wooding, S., Bamshad, M.J., Brassington, A.M.E., *et al.* (2003) Genetic variation among world populations: inferences from 100 *Alu* insertion polymorphisms. *Genome Research* **13**, 1607–1618.

Wetton, J.H., Tsang, C.S.F., Roney, C.A. and Spriggs, A.C. (2002) An extremely sensitive species specific ARMS PCR test for the presence of tiger bone DNA. *Forensic Science International* **126**, 137–144.

Whittaker, D.K. (1995) Forensic dentistry in the identification of victims and assailants. *Journal of Clinical Forensic Medicine* **2**, 145–151.

Wilkinson, C. (2004) *Forensic Facial Reconstruction*. Cambridge: Cambridge University Press.

Willey, P. and Heilman, A. (1987) Estimating time since death using plant roots and stems. *Journal of Forensic Sciences* **32**, 1264–1270.

Williams, M.C. (1996) Forensic examination of a mouse allegedly found in a previously sealed can of milk stout. *Forensic Science International* **82**, 211–215.

Willott, G.M. and Allard, J.E. (1982) Spermatozoa: their persistence after sexual intercourse. *Forensic Science International* **26**, 125–128.

Withrow, A.G., Sikorsky, J., Downs, J.C.U. and Fenger, T. (2003) Extraction and analysis of human nuclear and mitochondrial DNA from electron beam irradiated envelopes. *Journal of Forensic Sciences* **48**, 1302–1308.

Wyss, C. and Cherix, D. (2004) Murder followed by suicide in a forest: what could be learned from a comparative field experiment. *Proceedings of the 2nd Meeting of the European Association for Forensic Entomology*, London, 29–30 March 2004.

Zehner, R., Amendt, J. and Krettek, R. (2004a) STR typing of human DNA from fly larvae fed on decomposing bodies. *Journal of Forensic Sciences* **49**, 337–340.

Zehner, R., Amendt, J., Schutt, S., Sauer, J., Krettek, R. and Povolny, D. (2004b) Genetic identification of forensically important flesh flies (Diptera Sarcophagidae). *International Journal of Legal Medicine* **118**, 245–247.

Zuccato, E., Chiabrando, C., Castiglioni, S., Calamari, D., Bagnati, R., Schiarea, S. and Fanelli, R. (2005) Cocaine in surface waters: a new eividence-based tool to monitor community drug abuse. *Environmental Health: A Global Access Science* **14**, 4–14.

Index

Essential Forensic Biology, Alan Gunn
© 2006 John Wiley & Sons, Ltd